ORGANIC QUANTUM CHEMISTRY PROBLEMS

Rudolf Zahradník
and
Petr Čársky

The J. Heyrovský Institute of Physical Chemistry and Electrochemistry
Czechoslovak Academy of Sciences
Prague, Czechoslovakia

A PLENUM / ROSETTA EDITION

Library of Congress Cataloging in Publication Data

Zahradník, Rudolf.
  Organic quantum chemistry problems.

  "Plenum/Rosetta edition."
  Includes bibliographical references.
  1. Quantum chemistry — Problems, exercises, etc. 2. Chemistry, Organic — Problems, exercises, etc. I. Čársky, Petr, joint author. II. Title.
  [QD462.Z34 1973b]   547′.1′28   73-12931

ISBN 978-1-4684-8461-8   ISBN 978-1-4684-8459-5 (eBook)
DOI 10.1007/978-1-4684-8459-5

A Plenum/Rosetta Edition
Published by Plenum Publishing Corporation
227 West 17th Street, New York, N.Y. 10011

First paperback printing 1973

© 1972 Plenum Press, New York
A Division of Plenum Publishing Corporation

United Kingdom edition published by Plenum Press, London
A Division of Plenum Publishing Company, Ltd.
Davis House (4th Floor), 8 Scrubs Lane, Harlesden, London, NW10 6SE, England

INTRODUCTION

This collection of examples of the application of quantum-chemical methods in the study of various chemical problems (mainly in organic chemistry) arose from an arrangement of the materials used by the first of the authors in his lectures on organic quantum chemistry and chemical constitution, delivered since 1959 at the Science Department of the Charles University, as well as in courses of quantum chemistry and, finally, in his lectures on the fundamentals of quantum chemistry held in the years 1965-1967 at the universities of Würzburg, Groningen, and Giessen, and at the Technical University of Darmstadt.

This collection is to be regarded as a supplement to existing textbooks on organic quantum chemistry. Whereas the situation is satisfactory as far as the number and the standard of textbooks and monographs in this field are concerned, this cannot be said of the collections of examples. Although in some books[1-4] a certain number of examples are presented, we believe, in view of the great importance of numerical calculations, that a separate publication of examples would be useful. A similar subject has been dealt with in the recently published book by A. and O. Julg[5] and in the first part of a trilogy to be published by Heilbronner and Bock.[6]

In his work with this collection of examples, the student will undoubtedly utilize some of the standard textbooks, but nevertheless, we have summarized the quantum-chemical data in Part III, and presented some frequently employed quantum-chemical relationships in Part IV. Part I contains only the formulation of the problems, while Part II shows the procedures of the calculation and gives the results. The notes to the solutions (as well as the formulas in Part IV) are to be regarded only as aids facilitating the student's work, but not as a substitute for explanation or the study of a textbook.

Many of the examples are numerically laborious and thus time-consuming. It is conceivable that such examples cannot be presented in a lecture though they are of primary significance. The presence of these examples in the collection, together with the demonstration of the entire procedure, is also one of the reasons why the compilation of this book was undertaken.

The examples are divided into six groups according to their character. The first group relates to simple methods of quantum chemistry. Most of the examples of this group are tasks in the framework of the HMO method: establishment of the HMO matrix and calculation of orbital energies, expansion coefficients, and reactivity indices, i.e., basic HMO quantities, which are used in the subsequent groups of examples in chemical applications. Although in current practice the HMO calculations are carried out by means of computers, the HMO tasks in this collection are presented for numerical calculation. The solution of the problems in this way is time-consuming, laborious, and certainly of little interest from the chemical or mathematical viewpoint, but it is only their calculation that can appropriately acquaint the chemists with the HMO method. A similar situation is also encountered in the branch of more advanced methods of molecular orbitals, such as the ω-technique, the extended Hückel method, and the SCF-MO-LCAO, CNDO, and LCI-MO-LCAO methods. Although in these cases the calculations are still more extensive, some of them, e.g., numerical SCF calculations, have been included in the second group of examples. Since the calculations are rather lengthy, it is reasonable to demand only that the F-matrix be ascertained. One of the SCF examples is provided with the computer output to give the student an idea of the entire SCF procedure.

Whereas the examples of the first two groups are intended to acquaint the chemist with the conventional methods of molecular orbitals, the later groups are devoted to their applications. The examples in the third group pertain to electronic spectra, and the fourth group contains examples from the field of radio-frequency spectroscopy. The center of gravity of organic quantum chemistry is represented by the examples of the fifth group, because information about reactivity is most attractive to the chemist. Finally, the sixth group contains examples of a miscellaneous nature.

Despite all our efforts to eliminate errors, some may have escaped our attention. Our readers are requested, therefore, kindly to point out possible errors and shortcomings.

We wish to express our gratitude to Mrs. R. Žohová for her extensive and careful assistance in the preparation of the manuscript. Thanks are also due to Mrs. M. Zahradníková for reading the manuscript and galley proofs.

CONTENTS

NOTATION

a,b,c,d	atomic orbitals (VB method)
a_H	coupling constant (ESR)
A	atom localization energy
A	antisymmetric state
Å	angström (10^{-8} cm)
A_e	electrophilic atom localization energy
A_n	nucleophilic atom localization energy
A_o	ortho-localization energy
A_p	para-localization energy
A_r	radical atom localization energy
AO	atomic orbital
$c_{i\mu}$	expansion coefficient for the μ-th atomic orbital in the i-th molecular orbital
C	expansion coefficient of the VB and the LCI wave functions
D	Debye
E	HMO orbital energy (in β units); $E = \alpha - k\beta$; if $\alpha = 0$, then E (in β units) equals k
E_D	delocalization energy
$E_{D,m}$, $E_{D,n}$	specific delocalization energies (for C–C bond, for π-electron)
$E_{1/2}$	polarographic half-wave potential
$E(N \rightarrow V_1)$	HMO excitation energy for the transition of an electron from the HOMO into the LFMO
ESR	electron spin resonance
F	free valence
F	Hartree – Fock operator
$F_{\mu\nu}$	matrix elements of the Hartree–Fock operator
FE	free electron method
h	Planck constant

1

H_{KL}	resonance integral (VB)
HMO	simple MO-LCAO method (Hückel approximation, also for systems with heteroatoms)
HOMO	highest occupied molecular orbital
i,j	indices of molecular orbitals
J	simple resonance integral (VB)
k	$k = -(\alpha - E)/\beta$ (see E)
LCAO	linear combination of atomic orbitals
LCI	method of limited configuration interaction
LFMO	lowest free molecular orbital
MO	method of molecular orbitals
NBMO	nonbonding molecular orbital
NMR	nuclear magnetic resonance
NQR	nuclear quadrupole resonance
p	bond order
p	designation for the first intense band in the electronic spectra of benzenoid hydrocarbons
P	bond order
q	π-electron density
Q	Coulomb integral (VB)
Q	proportionality constant in the relation between the ESR coupling constant and spin density
$r_{\mu\nu}$	distance between the atoms μ and ν
S	superdelocalizability
S	overlap integral
S	symmetric state
S'	approximate superdelocalizability
SC	self-consistency (in the sence of the ω technique, etc.)
SCF	self-consistent field method
VB	valence-bond method
W	total HMO electron energy
α	see α_X
α	designation for a band in the electronic spectra of benzenoid hydrocarbons
α	spin symbol
α_X	Coulomb integral of the atomic orbital; X denotes the kind of atom, $\alpha_X = \alpha + \delta_X\beta$, where α is the Coulomb integral of the $2p_z$ atomic orbital of the carbon atom, β is the resonance integral of the π bond between two neighboring $2p_z$ orbitals, and δ_X is a constant

β	designation for a band in the electronic spectra of benzenoid hydrocarbons
β	spin symbol
β'	designation for a band in the electronic spectra of benzenoid hydrocarbons
β_{CX}	resonance integral of the CX bond; $\beta_{CX} = \rho_{CX}\,\beta$, ρ_{CX} being a constant; for β see α_X
γ	energy unit of the simple MO–LCAO method, if $S_{\mu\nu} = 0.25$ (μ and ν are neighboring atoms)
$\gamma_{\mu\nu}$	electron repulsion integral between the centers μ and ν
δ_X	see α_X
δ_{ij}	Kronecker delta
ε	molar extinction coefficient
ε_π	total SCF π-electron energy
χ_μ	μ-th atomic orbital
λ	constant in the McLachlan equation
λ	wavelength
μ	(as a subscript) designation of the atom
$\vec{\mu}$	dipole moment
$\vec{\mu}_\pi$	π-electron contribution to the total dipole moment
π	polarizability
ρ	spin density
ρ_{CX}	see β_{CX}
σ	plane of symmetry
φ_i	i-th molecular orbital
ψ, Ψ	wave function
ω	constant of the ω method

I: Exercises

1. SIMPLE METHODS

1.A. The Hückel Version of the MO − LCAO Method

1. Represent the angular part of the following atomic orbitals in the orthogonal coordinate system:

   (a) 1s, 2s
   (b) $2p_x$, $2p_y$, $2p_z$
   (c) all five d orbitals.

2. Represent (a) the σ-molecular orbitals of the $C-H$ bonds and the $C-C$ bond in ethylene as well as (b) both π-molecular orbitals.

3. Represent the overlap of the $2p_z$ orbitals in (2,2)-paracyclophane (I) and barrelene (II).

4. Represent the following orbitals (and overlap of orbitals) in benzene:

   (a) overlap of the atomic orbitals forming the $C-H$ and the σ $C-C$ bonds
   (b) localized orbitals of the σ bonds
   (c) all six delocalized π-molecular orbitals.

5. Represent all six delocalized π-molecular orbitals of fulvene. For determining the nodal planes employ the expansion coefficients (see p. 196).

6. Represent the energetically most favorable π-molecular orbital in (a) butadiene (for an idealized linear structure), (b) butadiine, (c) acrylonitrile, and (d) allene.

7. Represent the energetically most favorable and the energetically most unfavorable π-molecular orbitals for (a) butadiene, (b) hexatriene, (c) benzene, and (d) naphthalene. Give the characteristic properties of the energetically most favorable and the energetically most unfavorable molecular orbitals for these compounds.

8. (1) Represent the overlap of the atomic orbitals for pyridine, pyrrole, aniline, benzophenone, furan, and phenol.
    (a) Indicate the nonbonding atomic orbitals and the overlap of the atomic orbitals forming the σ bond.
    (b) Indicate the overlap of the atomic orbitals forming the π-molecular orbitals.

    (2) What is the difference between the lone pairs on oxygen in phenol and the free electron pairs on oxygen in benzophenone?

9. Explain without using molecular diagrams, directly on the basis of the structural formulas, why pyridine is a stronger base than pyrrole.

10. Ascertain the symmetry groups for the following compounds: (a) benzene, (b) cis-butadiene, (c) trans-butadiene, (d) triphenylene, (e) phenanthrene, (f) cyclopropenyl, (g) 2-aminonaphthalene, (h) cyclodecapentaene, (i) γ-pyrone, (j) pyrene, (k) diphenyl.

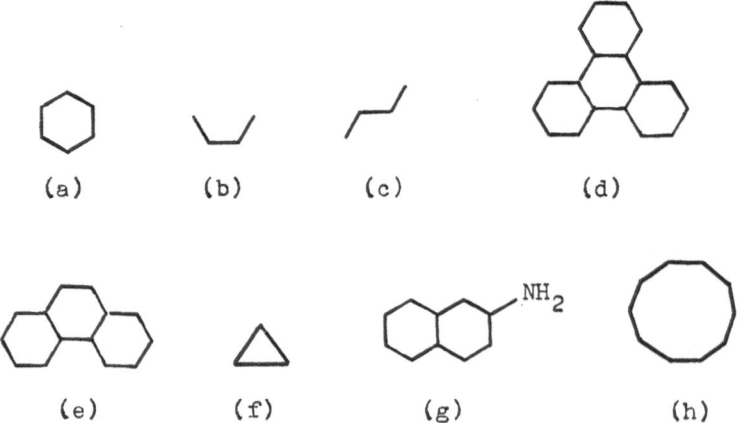

(a)   (b)   (c)   (d)

(e)   (f)   (g)   (h)

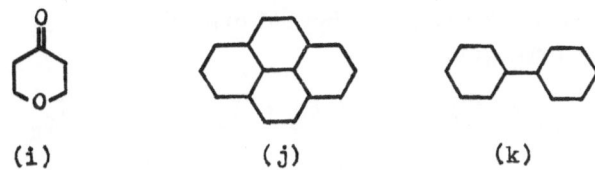

(i)   (j)   (k)

11. Indicate in the following group the alternant (A) and the nonalternant (N) hydrocarbons, as well as the even (E) and the odd (O) hydrocarbons:

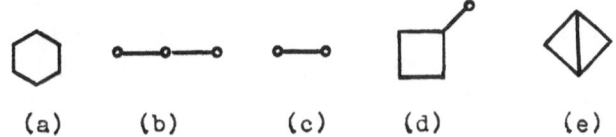

(a)   (b)   (c)   (d)   (e)

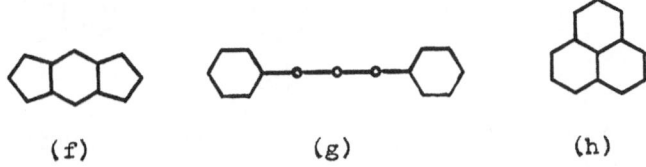

(f)   (g)   (h)

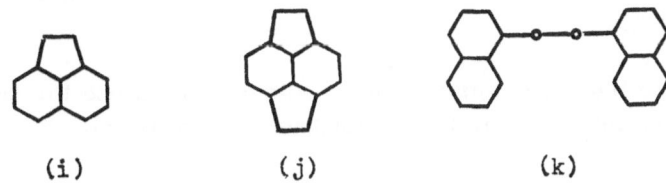

(i)   (j)   (k)

(l)   (m)   (n)

12. Prove that the carbon sp^3-hybrid orbitals* (φ_1, φ_2, φ_3, φ_4) are mutually orthogonal.[1a]

$$\varphi_1 = 1/2 \, (\chi_s + \chi_{p_x} + \chi_{p_y} + \chi_{p_z})$$

$$\varphi_2 = 1/2 \, (\chi_s - \chi_{p_x} - \chi_{p_y} + \chi_{p_z})$$

$$\varphi_3 = 1/2 \, (\chi_s + \chi_{p_x} - \chi_{p_y} - \chi_{p_z})$$

$$\varphi_4 = 1/2 \, (\chi_s - \chi_{p_x} + \chi_{p_y} - \chi_{p_z})$$

13. Set up the system of the secular equations, the secular determinant, and the HMO matrix for the following compounds: (a) butadiene, (b) ortho-quinodimethane, and (c) naphthalene. In case (c) utilize both planes of symmetry successively, and finally both planes of symmetry simultaneously.

$$CH_2 = CH - CH = CH_2$$

(a)   (b)   (c)

14. Set up and expand the secular determinant for (a) ethylene, (b) allyl, (c) butadiene, (d) cyclopropenyl, and (e) cyclobutadiene, and calculate the MO energies.

15. Establish the secular polynomial for 1,2-diazabutadiene ($\alpha_N = \alpha + 0.5\beta$).

16. Utilize the C_{2h} symmetry of 2,6-diazanaphthalene for setting up the system of secular equations ($\alpha_N = \alpha + 0.5\beta$).

---

*Note: χ_s, χ_{p_x}, χ_{p_y}, and χ_{p_z} are normalized and mutually orthogonal; consequently the following holds:

$$\int \chi_\mu^2 \, d\tau = 1 \; ; \qquad \int \chi_\mu \chi_\nu \, d\tau = 0 \, .$$

17. Set up the system of the secular equations, the secular determinant, and the HMO matrix for models of the following compounds: (a) 2-azabutadiene, (b) 1,2-thiazole, (c) para-benzoquinone. In case (c) utilize both planes of symmetry successively, and finally both planes of symmetry simultaneously.

Empirical parameters:

(a) $$\alpha_N = \alpha + 0.5\beta$$

(b) $$\alpha_N = \alpha + 0.5\beta$$
$$\alpha_S = \alpha + \beta$$
$$\beta_{CS} = \beta_{NS} = 0.6\beta$$

(c) $$\alpha_O = \alpha + 1.3\beta$$
$$\alpha_{C(O)^*} = \alpha + 0.2\beta$$
$$\beta_{CO} = \sqrt{2}\beta$$

$$CH_2{=}N{-}CH{=}CH_2$$

(a)   (b)   (c)

18. Examine whether the following secular determinant of fulvene is correct or incorrect. Give your arguments.

$$
\begin{vmatrix}
-k & 1 & 0 & 0 & 0 & 0 \\
1 & -k & 1 & 0 & 0 & 1 \\
0 & 1 & -k & 1 & 0 & 0 \\
0 & 0 & 1 & -k & 1 & 0 \\
0 & 0 & 0 & 1 & -k & 1 \\
0 & 0 & 0 & 0 & 1 & -k
\end{vmatrix}
$$

19. Set up the secular determinants for the anion, the radical, and the cation of cyclopentadienyl. In what do they differ?

---

*Coulomb integral of the carbon atom to which oxygen is bound.

20. Set up the secular equations for the calculation of the orbital ener-
    gies of thiophene for symmetric and antisymmetric states. Con-
    sider a model in which the participation of the d orbitals of sul-
    fur is not being taken into account, assuming the following para-
    meters:

	δ	ρ
(a)	0.0	0.6
(b)	0.5	0.6
(c)	1.0	0.6
(d)	1.0	0.8
(e)	1.0	1.0

21. Ascertain the HMO orbital energies of methylenecyclopropene by
    a graphical solution of its polynomial equation:

$$k^4 - 4k^2 - 2k + 1 = 0$$

22. Calculate the energy of the energetically most favorable π-mole-
    cular orbital of naphthalene when the orbital has the following
    form:

$$\varphi_1 = 0.300\,\chi_1 + 0.231\,\chi_2 + 0.231\,\chi_3 + 0.300\,\chi_4 +$$
$$+ 0.300\,\chi_5 + 0.231\,\chi_6 + 0.231\,\chi_7 + 0.300\,\chi_8 +$$
$$+ 0.461\,\chi_9 + 0.461\,\chi_{10}$$

23. Calculate the orbital energies of 1-azabutadiene and 1-aminobu-
    tadiene without expansion of the secular determinants. For the
    perturbation calculation employ the orbital energies and expansion
    coefficients of the MO's of butadiene and pentadienyl (see pp. 191,
    192). For the $2p_z$ orbitals of the nitrogen atoms use the following
    parameters:

$$\alpha_N = \alpha + 0.5\beta \qquad\qquad \alpha_N = \alpha + \beta$$

  (for nitrogen of   (for nitrogen of
   the type $-\bar{N}=$)   the type $-\bar{N}H_2$)

24. Determine the energies of three molecular orbitals of m-quinodi-
    methane (I) (without solving the secular equations). For this pur-
    pose employ the orbital energies for the allyl radical; calculate
    its orbital energy likewise without solving the secular equations.

I

25. Try to utilize the known orbital energies of benzene for the deter-
    mination of six orbital energies for anthracene, 9-aminoanthracene,
    and 9, 10-diaminoanthracene.

26. Assume that you know four orbital energies of fulvene ($E_1 = \alpha +$
    2.115β, $E_2 = \alpha + 0.618\beta$, $E_3 = \alpha - 1.618\beta$, $E_4 = \alpha - 1.861\beta$); deter-
    mine the remaining two energies without solving the polynomial
    equation.

27. (a) On the basis of the structural formula of 2-vinylallyl and the
    knowledge of the energy of one molecular orbital ($k_1 = E = \alpha +$
    0.765β), suggest a way for determining the remaining four ener-
    gies without expanding the secular determinant.

    (b) The energy of the bonding molecular orbital of the allyl radical
    is $\alpha + \sqrt{2}\beta$. Calculate all seven orbital energies for the given
    hypothetic compound without solving the secular determinant.

28. Calculate the orbital energies, the expansion coefficients, the bond
    orders, and the total π-electron energies for the cation, the radical,
    and the anion of the allyl system.

29. Calculate the expansion coefficients of the lowest bonding MO of benzyl by employing Cramer's rule (solution of a system of linear equations by means of determinants) when you know the orbital energy $E = \alpha + 2.101\beta$ and the secular equations for the symmetric and antisymmetric states.

Sx:

$$-kc_1 + c_2 \qquad\qquad = 0$$
$$c_1 - kc_2 + 2c_3 = 0$$
$$c_2 - kc_3 + c_4 = 0$$
$$c_3 - kc_4 + c_5 = 0$$
$$2c_4 - kc_5 \qquad\qquad = 0$$

Ax:

$$-kc_3 + c_4 = 0$$
$$-kc_4 + c_3 = 0$$

30. The following tables contain the energies and expansion coefficients of the bonding π-molecular orbitals of butadiene, p-quinodimethane, and naphthalene. Supplement these data for the antibonding orbitals. Make sure (a) that the two orbitals richest in energy are orthogonal and (b) that (for the C_1 atoms) the sum of the squares of the expansion coefficients amounts to 0.500 for the antibonding orbitals.

Butadiene

	$E(\beta)$	Sym.	c_1	c_2	c_3	c_4
φ_1	1.61803	Sx	0.37175	0.60150	0.60150	0.37175
φ_2	0.61803	Ax	0.60150	0.37175	-0.37175	-0.60150
φ_3						
φ_4						

p–Quinodimethane

	E(β)	Sym.	c_1	c_2	c_3	c_4
φ_1	2.17009	SxSy	0.43249	0.36962	0.36962	0.43249
φ_2	1.48119	SxAy	0.52990	0.21357	−0.21357	−0.52990
φ_3	1.00000	AxSy	0.00000	0.50000	0.50000	0.00000
φ_4	0.31111	SxSy	0.17934	−0.26033	−0.26033	0.17934
φ_5						
φ_6						
φ_7						
φ_8						

p–Quinodimethane (cont.)

	E(β)	Sym.	c_5	c_6	c_7	c_8
φ_1	2.17009	SxSy	0.36962	0.36962	0.19929	0.19929
φ_2	1.48119	SxAy	−0.21357	0.21357	0.35775	−0.35775
φ_3	1.00000	AxSy	−0.50000	−0.50000	0.00000	0.00000
φ_4	0.31111	SxSy	−0.26033	−0.26033	0.57645	0.57645
φ_5						
φ_6						
φ_7						
φ_8						

Naphthalene

	$E(\beta)$	Sym.	c_1	c_2	c_3	c_4	c_5
φ_1	2.30278	SxSy	0.30055	0.23070	0.23070	0.30055	0.30055
φ_2	1.61803	SxAy	0.26287	0.42533	0.42533	0.26287	−0.26287
φ_3	1.30278	AxSy	0.39958	0.17352	−0.17352	−0.39958	−0.39958
φ_4	1.00000	SxSy	0.00000	−0.40825	−0.40825	0.00000	0.00000
φ_5	0.61803	AxAy	0.42533	0.26286	−0.26286	−0.42533	−0.42533
φ_6							
φ_7							
φ_8							
φ_9							
φ_{10}							

Naphthalene (cont.)

	$E(\beta)$	Sym.	c_6	c_7	c_8	c_9	c_{10}
φ_1	2.30278	SxSy	0.23070	0.23070	0.30055	0.46140	0.46140
φ_2	1.61803	SxAy	−0.42533	−0.42533	−0.26287	0.00000	0.00000
φ_3	1.30278	AxSy	−0.17352	0.17352	0.39958	0.34705	−0.34705
φ_4	1.00000	SxSy	−0.40825	−0.40825	0.00000	0.40825	0.40825
φ_5	0.61803	AxAy	0.26286	−0.26286	−0.42533	0.00000	0.00000
φ_6							
φ_7							
φ_8							
φ_9							
φ_{10}							

31. Calculate the total π-electron energy of ethylene (a) in the ground state and (b) in the $N \rightarrow V_1$ excited state, and (c) calculate the total π-electron energy of the dianion $C_2H_4^{2-}$.

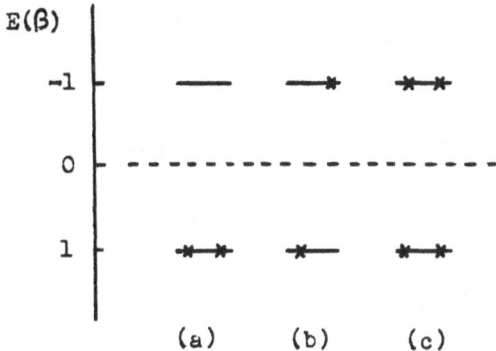

(a)   (b)   (c)

32. (a) Is it possible to calculate the total π-electron energy of fulvene from the orbital energies of its antibonding MO's?

(b) Is it possible to calculate the total π-electron energy of pyridine from the analogous quantities?

(c) Is it possible to calculate the energies of the bonding MO's in general from the known values of the orbital energies of the antibonding MO's?

33. Naphthalene (I) and azulene (II) have the following orbital energies [given in the form $-k = (\alpha - E)/\beta$].

I: 2.303; 1.618; 1.303; 1.000; 0.618; −0.618; −1.000; −1.303; −1.618; −2.303

II: 2.310; 1.652; 1.356; 0.887; 0.477; −0.400; −0.738; −1.579; −1.869; −2.095

Calculate the total π-electron energies, delocalization energies, specific delocalization energies (delocalization energies related to one π-electron or one C−C bond), and the energies of the $N \rightarrow V_1$ transitions.

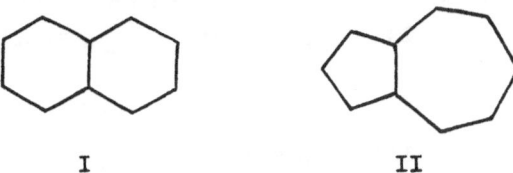

I   II

34. To the compounds pyrrole (I), furan (II), and thiophene (III) correspond the following values of orbital energies* in the form $-k = (\alpha - E)/\beta$.

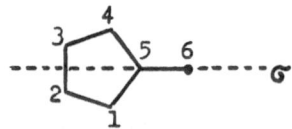

I   II   III

I:	2.230;	1.189;	0.618;	-1.008;	-1.618
II:	2.633;	1.314;	0.618;	-0.948;	-1.618
III:	2.091;	1.000;	0.618;	-1.091;	-1.618

Calculate the total π-electron energies, delocalization energies, specific delocalization energies, and the energies of the $N \rightarrow V_1$ transitions.

35. The orbital energies [in the form $-k = (\alpha - E)/\beta$] of the seven occupied orbitals of anthracene amount to 2.414; 2.000; 1.414; 1.414; 1.000; 1.000; 0.414.

    (a) How high are the energies of the unoccupied MO's?
    (b) How high is the total π-electron energy?
    (c) How high is the energy of the $N \rightarrow V_1$ transition?
    (d) What is the value of the E_D in β and in kcal/mole (let us assume that in this particular case β = 20 kcal/mole)?

36. Calculate the total π-electron energy of fulvene (I) from the known energies of the respective symmetric bonding MO's ($\alpha + 2.1149\beta$; $\alpha + \beta$) and from the known coefficients c_1 and c_2 of the antisymmetric bonding MO ($c_1 = 0.6015$; $c_2 = 0.3718$), without solving the secular equations for the antisymmetric states.

*The calculation was carried out for the following parameters:

I: δ_N = 1.5; φ_{C-N} = 0.8

II: δ_O = 2.0; φ_{C-O} = 0.8

III: δ_S = 1.0; φ_{C-S} = 0.8

37. How high is the π-electron density of pentalene in position 1 ?

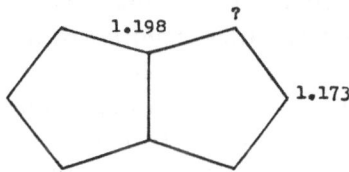

38. Assume that only the data given in the molecular diagram are at
your disposal. On the assumption that the total π-electron energy of fulvene equals $6\alpha + 7.466\beta$, calculate the missing quantities
q and p.

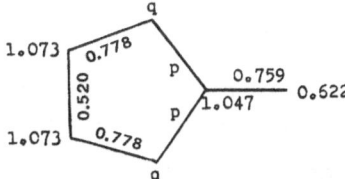

39. What is the magnitude of the π-bond order $9-10$ in naphthalene ?
The other bond orders are

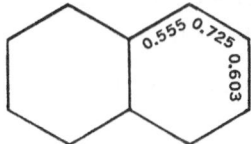

The total π-electron energy of naphthalene is $10\alpha + 13.683\beta$. Calculate the delocalization energy of naphthalene.

40. Assume that only the expansion coefficients of the antibonding molecular orbitals of pyridine are at your disposal (see p. 198). Calculate the π-electron density and the bond orders. Is it possible
to calculate the superdelocalizabilities for electrophilic substitution from these data ?

41. Calculate the molecular diagrams for methylenecyclopropene in
the excited states arising from $1 \rightarrow 1'$ and $1 \rightarrow 2'$ excitations. Check
the ascertained bond orders and electron densities by means of the
total π-electron energy. The expansion coefficients and orbital
energies are given on p. 196.

42. The values of the orbital energies and expansion coefficients of the
π-molecular orbitals of naphthalene were obtained by the solution
of the secular equations (the results are presented in the table on
p. 193). Verify whether the π-molecular orbitals are normalized.
Then use the values of the table to calculate the electron density
in position 1, the $1-2$ and $1-9$ bond orders, the free valence, and
the approximate and the accurate superdelocalizability in position 1.

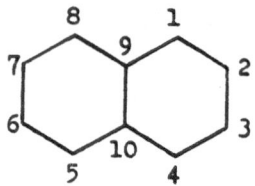

43. Use the values of the expansion coefficients of the MO's of fulvene
and pyridine (see pp. 196, 198) to calculate:

    (a) the π-electron densities
    (b) the bond orders
    (c) the free valences
    (d) the approximate superdelocalizabilities
    (e) the accurate superdelocalizabilities (d and e for both radical
        and polar substitutions).

    Check the quantities a, b, c, d, e.

1.B.   Free Electron Method (FEMO)

44. Draw schematically the HMO and the FEMO diagrams of orbital
levels of octatetraene and compare them. (For application of the
FEMO method, see also Exercise 71).

1.C.   Valence Bond Method (VB)

45. By means of the VB method calculate the wave function, the π-elec-
tron energy, and the delocalization energy of butadiene.

46. By means of the VB method calculate the wave function, the π-elec-
tron energy, and the delocalization energy of benzene.

47. Draw the complete system of the canonical structures for ben-
    zocyclobutadiene.

48. The application of the VB method to more extensive conjugated
    systems is difficult since, because of the large number of canonical
    structures, the calculations are very laborious. Determine how
    many structures would be necessary to ascertain the wave func-
    tions of benzene, naphthalene, anthracene, and tetracene if the for-
    mula for the number of canonical structures is $(2t)!/t!\,(t+1)!$,
    where $(2t)$ denotes the number of atomic orbitals in conjugation.

1.D. Perturbation Treatment and Special Methods

49. The values of the atom—atom polarizabilities for the C_1 atom of
    naphthalene are as follows:

μ	1	2	3	4	5	6
$\pi_{1,\mu}[\beta^{-1}]$	0.443	−0.213	0.018	−0.139	−0.023	0.007

μ	7	8	9	10
$\pi_{1,\mu}[\beta^{-1}]$	−0.033	0.027	−0.089	0.004

By means of these data calculate the electron densities of the
quinoline model ($\alpha_N = \alpha + 0.5\beta$). Check the calculation.

50. Use equation (36) of the Appendix

$$\sum_i \frac{c_{i\mu}^2}{E - k_i} = \frac{E - \delta_\nu}{\varrho_{\mu\nu}^2 + (E - \delta_\nu)\delta_\mu} \qquad (36)$$

and the knowledge of the orbital energies and expansion coefficients
(HMO) of methylenecyclopropene (I) (p. 196) to calculate (a) the
localization energy (A_n) in position μ (I) and (b) the orbital energies
E of the model of the heteroanalog of methylenecyclopropene (II)
and of a derivative of methylenecyclopropene, without (III) and with
(IV) regard to the inductive effect.

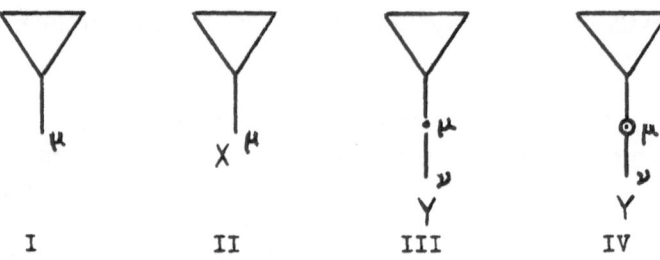

I   II   III   IV

Empirical parameters:

$$\alpha_\mu = \alpha + \delta_\mu \beta$$

$$\beta_{\mu\nu} = \varsigma_{\mu\nu} \beta$$

$\delta_\mu = 0.5$ $\delta_\mu = 0; \delta_\nu = 1$ $\delta_\mu = 0.2; \delta_\nu = 1$

                $\varsigma_{\mu\nu} = 1$ $\varsigma_{\mu\nu} = 0.7$

II   III   IV

51. Calculate the values of the atom−atom, atom−bond, bond−atom, and bond−bond polarizabilities in the methylenecyclopropene molecule. Give the results exact to three decimal places and summarize them in a table. The necessary values of the orbital energies and expansion coefficients of the π-molecular orbitals are tabulated on page 196. Check the calculated values.

52. Calculate the π-electron densities of the benzyl cation without solving the secular equations.

53. Indicate the positions where the electron density decreases most intensely when nitrogen is introduced as a heteroatom into position 1 or 2 in phenanthrene.

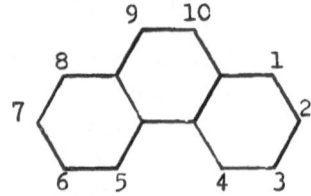

54. Employ the perturbation calculation for determining the π-electron densities and bond orders for 1, 3-thiazole from the data for benzene.

Hint: First form thiophene by means of the Longuet-Higgins model[7]

(the resonance integrals of the weakened bonds 1,2 and 5,6 in benzene have the value 0.6β); then convert benzene into pyridine ($\alpha_N = \alpha + 0.5\beta$). Regard the superposition of the two results as a model of 1,3-thiazole.

The bond order in benzene is 0.666.

Atom—atom polarizabilities of benzene:

	1	2	3	4	5	6
1	0.398					
2	−0.157	0.398				
3	0.009	−0.157	0.398			
4	−0.102	0.009	−0.157	0.398		
5	0.009	−0.102	0.009	−0.157	0.398	
6	−0.157	0.009	−0.102	0.009	−0.157	0.398

Bond—bond polarizabilities:

	1-2	2-3	3-4	1-6	5-6	4-5
1-2	0.241					
2-3	−0.204	0.241				
3-4	0.130	−0.204	0.241			
1-6	−0.204	0.130	−0.092	0.241		
5-6	0.130	−0.092	0.130	−0.204	0.241	
4-5	−0.092	0.130	−0.204	0.130	−0.204	0.241

Compare the results with the molecular diagram obtained by solution of the secular equations with the use of the same parameters

$(\rho_{C-S} = 0.6; \delta_N = 0.5)$:

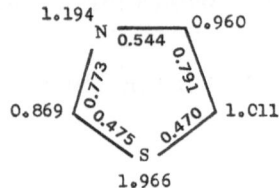

2. ADVANCED METHODS

2.A. ω Technique, Related SC Methods and Extended Hückel Method

55. Ascertain the dependence of the π-electron densities of the allyl cation on the number of approximations in the ω technique ($\omega = 1.4$). (According to Ref. 1a).

56. Set up the matrices according to Janssen and Sandström (simultaneous α- and β-variation) for acrolein and urea for the following set (A) of parameters:

(A) $\delta_O = 1$ $\rho_{CO} = 1.2$ $\omega = 1.0$
            $\delta_N = 1.5$ $\rho_{CN} = 1.2$

The calculated π-electron densities, bond orders, and orbital energies (HMO and iterative values) are tabulated on page 23. Discuss the results and take also into consideration the second set of HMO data based on modified parameters (B):

(B) $\rho_{CO} = 1.8$ (all other values remain unchanged)

57. Ascertain the matrix according to Coulson and Golebiewski for methylenecyclopropene. The molecular diagram is given on p. 204.

58. Set up the matrix of the effective Hamiltonian and the overlap matrix in the method according to Hoffmann[8] for methylene CH_2. Assume the following geometry: 1.1 Å for the $C-H$ bond length and 120° for the $H-C-H$ bond angle. For the diagonal matrix elements

3

N

>C—O

N   2 4

1

O—C—C—O

1 2 3 4

	HMO(A)	HMO(B)	SC	HMO(A)	HMO(B)	SC
q_1	0.813	0.892	0.947	1.796	1.850	1.821
q_2	1.035	1.032	0.998	0.702	0.764	0.842
q_3	0.699	0.771	0.846	1.796	1.850	1.821
q_4	1.453	1.305	1.210	1.707	1.537	1.516
p_{12}	0.886	0.931	0.934	0.515	0.431	0.455
p_{23}	0.453	0.357	0.354	0.515	0.431	0.455
$p_{34}(p_{24})$	0.810	0.895	0.917	0.617	0.757	0.749
k_1	2.036	2.551	2.615	2.871	3.183	3.478
k_2	1.000	1.000	1.346	1.500	1.500	1.679
k_3	-0.444	-0.678	-0.840	1.153	1.252	0.909
k_4	-1.592	-1.874	-2.121	-1.523	-1.940	-2.067

use the following values (in eV):

$$H_{\mu\mu} (C\ 2p) = -11.4$$
$$H_{\mu\mu} (C\ 2s) = -21.4$$
$$H_{\mu\mu} (H\ 1s) = -13.6$$

Calculate the nondiagonal matrix elements by means of the following expression:

$$H_{\mu\nu} = 0.5K\ (H_{\mu\mu} + H_{\nu\nu})\ S_{\mu\nu} ,$$

where $K = 1.75$, $H_{\mu\mu}$ ($H_{\nu\nu}$) stands for the diagonal matrix element
of the μ-th (ν-th) AO, and S denotes the overlap integral between
the μ-th and ν-th AO's. The values of the latter are available in
the literature.[9]

(Hint: The tabulated values of $S_{\mu\nu}$ concern biatomic molecules. Therefore, we have to use projections of the $2p_x$ and $2p_y$ AO's in the directions of the C – H bonds instead of these orbitals as such. This means that the overlap integral belonging to the distance 1.1 Å has to be multiplied by cosine of the respective angle.)

2.B. Self-Consistent Field Method (SCF-MO-LCAO)

59. (a) Construct the F matrix according to Pople for trans-butadiene. Use the HMO expansion coefficients (see p. 191). As a starting set for the geometric conditions assume all bonds to be of equal length ($r_{12} = r_{23} = r_{34} = 1.4$ Å) and the angles of the C – C bonds to be equal to 120°. Approximate the electronic repulsion integrals according to Mataga – Nishimoto (see Appendix). For the resonance integrals between neighboring orbitals use – 2.318 eV; for the valence state ionization potential of the carbon atom use the value of – 11.22 eV.

(b) Employ the bond order matrix from the eighth SCF iteration, set up the F matrix, and decide whether "self-consistency" has already been attained.

F Matrix (in eV) after the seventh SCF iteration:

	1	2	3	4
1	–5.95500	–4.72957	0.00000	0.53243
2		–5.95500	–3.29372	0.00000
3			–5.95500	–4.72957
4				–5.95500

Bond order matrix, $p_{\mu\nu}$, the eighth iteration:

	1	2	3	4
1	1.00000	0.92703	0.00000	–0.37498
2		1.00000	0.37498	0.00000
3			1.00000	0.92703
4				1.00000

60. Calculate the total SCF π-electron energy of naphthalene when you know the SCF bond orders and the interatomic distances. Ap-

proximate the repulsion integrals according to Mataga – Nishimoto:

$$\gamma_{\mu\nu} = \frac{14.399}{1.367 + r_{\mu\nu}} .$$

The calculation was carried out with the value $\beta_{\mu\nu} = -2.39$ eV.

(Note: The value for β used in this example differs somewhat from those given in Exercise 59.)

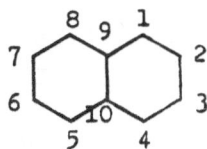

Interatomic distances $r_{\mu\nu}$ (Å):

	1	2	3	4	5	6	7	8	9	10
1	–	1.400	2.425	2.800	3.704	4.200	3.704	2.425	1.400	2.425
2			1.400	2.425	4.200	5.048	4.850	3.704	2.425	2.800
3				1.400	3.704	4.850	5.048	4.200	2.800	2.425
4					2.425	3.704	4.200	3.704	2.425	1.400
5						1.400	2.425	2.800	2.425	1.400
6							1.400	2.425	2.800	2.425
7								1.400	2.425	2.800
8									1.400	2.425
9										1.400
10										–

Matrix of the bond orders $p_{\mu\nu}$:

	1	2	3	4	5	6	7	8	9	10
1	1.000	0.741	0.000	–0.358	0.086	0.000	–0.155	0.000	0.540	0.000
2		1.000	0.587	0.000	0.000	0.143	0.000	–0.155	0.000	–0.249
3			1.000	0.741	–0.155	0.000	0.143	0.000	–0.249	0.000
4				1.000	0.000	–0.155	0.000	0.086	0.000	0.540
5					1.000	0.741	0.000	–0.358	0.000	0.540
6						1.000	0.587	0.000	–0.249	0.000
7							1.000	0.741	0.000	–0.249
8								1.000	0.540	0.000
9									1.000	0.541
10										1.000

61. Calculate the following SCF (Pople) characteristics of 1,3-trans-butadiene from the known values of molecular integrals:

(a) Orbital energies E_1, E_2, and E_3

$$(E_i = H_{ii}^c + \sum_j^{occ} (2 J_{ij} - K_{ij}))$$

(b) Total SCF electronic energy ε (neglect the core – core repulsions). Make sure that this energy is not equal to double the sum of orbital energies

$$(\varepsilon = 2 \sum_i^{occ} H_{ii}^c + \sum_i^{occ} \sum_j^{occ} (2 J_{ij} - K_{ij}))$$

(c) Excitation energy of the singlet – singlet and singlet – triplet $\varphi_1^2\varphi_2^2 \rightarrow \varphi_1^2\varphi_2\varphi_3$ transitions. On what does the singlet – triplet splitting depend? Why is the singlet – triplet splitting very small with $n \rightarrow \pi^*$ transitions?

(d) By means of recalculation make sure that the values of the J_{11}, J_{12}, K_{12}, and H_{11}^c integrals are correct. For SCF expansion coefficients and repulsion γ integrals, see Exercise 59.

Molecular integrals (eV):

$$H_{11}^c = -28.426 \qquad J_{11} = 6.304 \qquad K_{12} = 1.701$$
$$H_{22}^c = -25.222 \qquad J_{22} = 5.861 \qquad K_{13} = 1.193$$
$$H_{33}^c = -22.373 \qquad J_{33} = 5.861 \qquad K_{23} = 1.775$$
$$J_{12} = 5.693$$
$$J_{13} = 5.693$$
$$J_{23} = 5.861$$

62. Derive the expression for the dependence of the Coulomb repulsion J_{mm} integral* (m in the index of the singly occupied MO) on the

---

*The integrals of this type are important in the MO approach to radicals (total energies, ionization potentials, disproportionation equilibria; cf. R. Zahradník and P. Čársky, Prog. Phys. Org. Chem., in press).

length of the chain in odd polyene radicals (I)

$$CH_2 \cdots \left(CH \cdots CH \right)_{\overline{i}} H$$

I (i = 0, 1, 2, ...)

In the LCAO expansion take into account only repulsion integrals for one center ($\gamma_{\mu\mu}$ = 10.53 eV) and neighboring centers ($\gamma_{\mu\nu}$ = 5.20 eV).

63. In the starting step of the CNDO/2 calculation [J. A. Pople and G. A. Segal, J. Chem. Phys. 44:3289 (1966)] a Hückel-type matrix is set up. Do this for hydrogen fluoride. The values (in eV) of $\frac{1}{2}$(I + A) and bonding β_A^o parameters necessary for the calculation are:

	Hydrogen	Fluorine
β_A^o	-9.0	-39.0
$\frac{1}{2}(I_s + A_s)$	7.176	32.272
$\frac{1}{2}(I_p + A_p)$	-	11.080

If the interatomic distance H−F is considered to be 0.9175 Å and oriented in the x direction, the following overlap integrals are obtained:

	H(1s)	F(2s)	F($2p_x$)	F($2p_y$)	F($2p_z$)
H(1s)	1	0.4490	0.3339	0	0
F(2s)		1	0	0	0
F($2p_x$)			1	0	0
F($2p_y$)				1	0
F($2p_z$)					1

64. The Hückel-type matrix (from Exercise 63) yields the following initial set of molecular orbital coefficients $c_{i\mu}$:

	1	2	3	4	5
	−36.5889	−15.3513	−11.0800	−11.0800	1.4123
H(1s)	0.36939	0.45056	0	0	0.81274
F(2s)	0.92200	−0.28692	0	0	−0.25999
F(2p_x)	0.11605	0.84538	0	0	−0.52140
F(2p_y)	0	0	1.00000	0	0
F(2p_z)	0	0	0	1.00000	0

Calculated electron repulsion integrals, γ_{AB}, have the following values: $\gamma_{HH} = 20.407$ eV, $\gamma_{HF} = 14.257$ eV, and $\gamma_{FF} = 25.700$ eV. From these data calculate the core matrix elements $H_{\mu\nu}$ and the matrix elements $F_{\mu\nu}$ for the first SCF iteration in the CNDO/2 method.

2.C.   Limited Configuration Interaction (LCI-MO-LCAO)

65. Employ the method of limited configuration interaction (Pariser − Parr) to calculate the excitation energies and the corresponding LCI wave functions for trans-butadiene. Use the HMO expansion coefficients as a starting set and consider the interaction between all four possible singly excited states. Carry out the calculation for the idealized geometry of the molecule ($r_{12} = r_{23} = r_{34} = 1.4$ Å, C − C − C bond angles equal to 120°). For the valence state ionization potential of the carbon atom use the value of − 11.22 eV and approximate the electron repulsion integrals according to Mataga and Nishimoto (see Appendix). Assume the resonance integrals between neighboring orbitals to be equal to − 2.318 eV.

3.   ELECTRONIC SPECTRA

66. For the hydrocarbons whose schematic formulas are given below, calculate the energies of the N → V_1 transitions without taking the

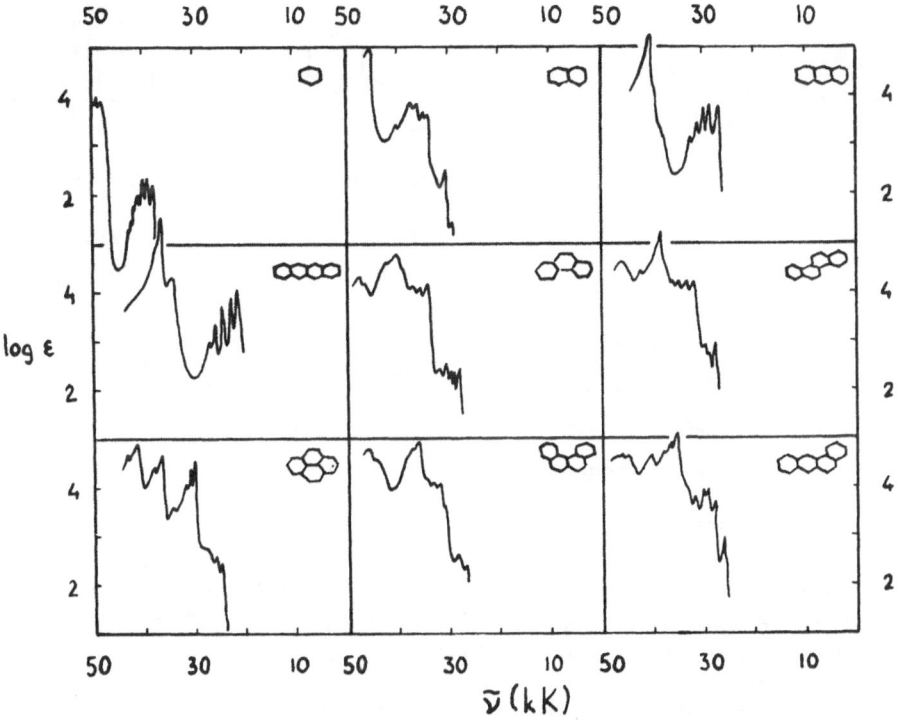

Fig. 1. Electronic spectra of benzenoid hydrocarbons.

overlap* into account ($S_{\mu\nu} = 0$ for $\mu \neq \nu$) and also when the overlap is taken into account. Assume the overlap integral between neighboring atoms to have the uniform value $S_{\mu\nu} = 0.25$ (μ and ν are neighboring atoms). Tabulate the results.

Benzene	Naphthalene	Anthracene	Tetracene
I	II	III	IV

---

* Take these values from the table on p. 202.

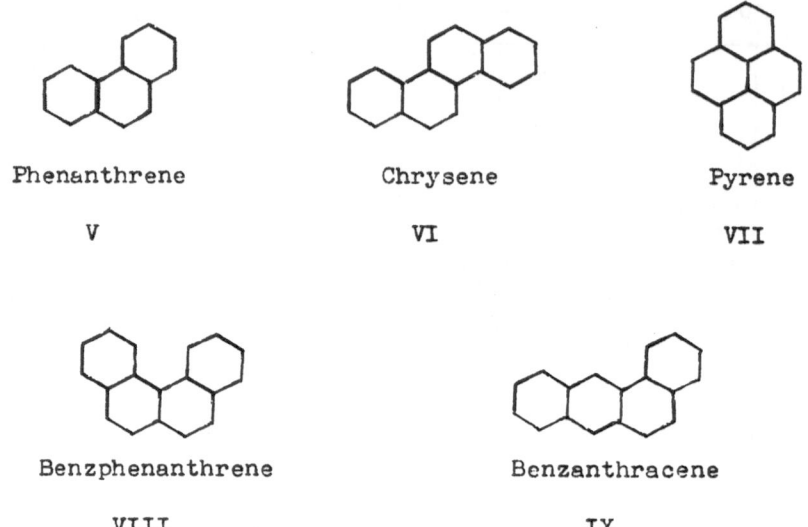

Phenanthrene   Chrysene   Pyrene

V   VI   VII

Benzphenanthrene   Benzanthracene

VIII   IX

Assign the α, p, β, and β' bands to the absorption curves of the hydrocarbons I–IX (Fig. 1); tabulate the positions of the maxima of the bands in mμ and in kK (10^3 cm^{-1}).

Ascertain graphically the dependence of the experimental wave numbers of the maxima of the p bands on the energies $E(N \rightarrow V_1)$ (in β and γ units). Note that the diagram of the linear dependence does not pass through zero; what is the probable cause of this fact?

67. Ascertain whether for the aromatic hydrocarbons I–IX of the preceding example there exists a statistically significant dependence between the wave number of the p bands (in kK) and the energy of the $N \rightarrow V_1$ transitions (in β units). Draw a regression line through the points and calculate the correlation coefficient. Further, verify that the energies of the $N \rightarrow V_1$ transitions in β and in γ units are mutually interrelated; discuss this dependence.

68. An LCI–SCF calculation of the position of the first band in the electronic spectrum of isobenzofulvene (I)* (which is due to a nearly

---

*This hydrocarbon has not yet been prepared; attempts to synthesize the amino derivatives II and III were successful, however.[10]

pure $1 \rightarrow 1'$ transition)

I   II   III

yields the value 17.3 kK. This wave number corresponds to the blue-violet color of I. Suggest an explanation why the amino derivative II is yellow (with λ_{max} = 376 mμ) by means of the first-order perturbation treatment. Take only the inductive effect of the amino group into consideration. What could be the reason for the blue color of its diphenyl derivative III (λ_{max} = 593 mμ) ? (Consider only the steric effect of the phenyl groups.)

The values of the expansion coefficient in position 10 (exocyclic carbon) in the lowest free and the highest occupied molecular orbitals are as follows:

$$c_{6,10} = 0.739; \qquad c_{5,10} = 0.000$$

The change in the Coulomb integral[3] in the substituted position due to the amino group amounts to 13.8 kK.

69. Try to interpret the electronic absorption spectrum (Fig. 2) of acenaphthylene[12] (I) in terms of results of the LCI-SCF calculation (see table on p. 32). Besides the absorption curve (in ethanol), Fig.

I

2 also gives the fluorescence polarization spectrum. The maxima and minima on this curve correspond to electronic transitions which are mutually perpendicularly polarized. (For more detailed information, see Ref. 12, from which these data are taken.)

LCI wave functions (ψ_i), their energies (E_i), oscillator strengths ($f_{o \to j}$) and cosine of the angle θ between the direction of polarization and the positive part of the x axis. (Calculation was performed with 25 singly excited configurations; γ integrals were approximated by the Mataga– Nishimoto formula)

ψ_i	E_i (eV)	$f_{o \to j}$	cos θ
ψ_0	0		
ψ_1	2.75	0.04	1
ψ_2	3.75	0.31	0
ψ_3	3.83	0.22	1
ψ_4	4.75	0.001	0
ψ_5	5.56	1.21	1

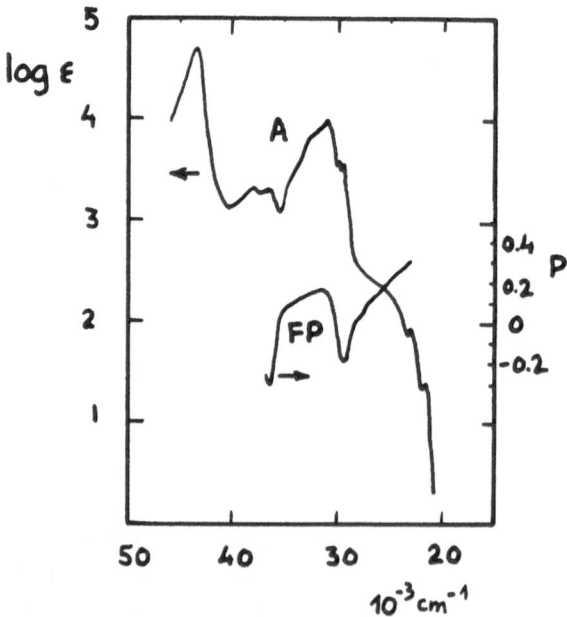

Fig. 2. Electronic spectrum of acenaphthylene. A; absorption curve, FP; fluorescence polarization curve. [12]

The coefficient $C_{i,j \to k}$ of the main configurations in the relation

$$\Psi_i = c_{i,o} \Delta_o + \sum_{jk} c_{i,j \to k} \Delta_{j \to k}$$

are as follows (Δ_o stands for the Slater determinant of the ground state, $\Delta_{j \to k}$ means a linear combination of Slater determinants for a monoexcited configuration formed by promoting an electron from the j-th to the k-th molecular orbital):

Ψ_o .... Δ_o (ground state configuration)

Ψ_1 .... $0.954 \Delta_{1 \to 1'}$

Ψ_2 .... $0.927 \Delta_{2 \to 1'}$

Ψ_3 .... $-0.500 \Delta_{2 \to 2'}$, $0.805 \Delta_{3 \to 1'}$

Ψ_4 .... $0.965 \Delta_{1 \to 2'}$

Ψ_5 .... $-0.308 \Delta_{1 \to 3'}$, $0.713 \Delta_{2 \to 2'}$, $0.355 \Delta_{3 \to 1'}$,
$0.402 \Delta_{5 \to 1'}$

The MO's of acenaphthylene are designed as follows:

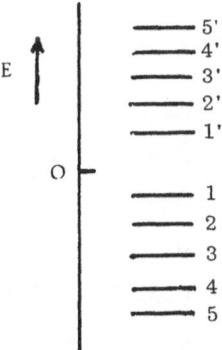

Hint: Introduce the LCI data into the graph of the absorption curve. convert $\log f_{o \to j}$ into $\log \varepsilon$ by means of an empirical relation

$$\log \varepsilon = \log f + 4$$

70. Interpret the electronic spectrum of aniline (Fig. 3) with the aid of the LCI-SCF (Pariser – Parr – Pople) calculation. The ground and excited state functions can be constructed from the SCF expansion coefficients (see table). Carry out the configuration interaction between all singly excited states arising from transitions with HMO energy less than or equal to 2.2β. Consider an idealized molecule geometry having all bond lengths equal to 1.40 Å and the $C – C – C$ and $C – C – N$ angles equal to 120°. The electronic repulsion integrals $\gamma_{\mu\nu}$ can be evaluated with the aid of an approximation due to Mataga and Nishimoto (see Appendix); if one of the centers μ or ν is the nitrogen atom, employ the formula

$$\gamma_{\mu\nu} = \frac{14.399}{1.001 + r_{\mu\nu}}$$

For the one-center γ_{NN} integral take the value 18.0 eV. Use the following semiempirical parameters: $\beta_{CC} = - 2.318$ eV, $\beta_{CN} = - 1.910$ eV, $I(C) = - 11.22$ eV, $I(N) = -27.3$ eV.

SCF orbital energies and expansion coefficients:

E (eV)	Symmetry	c_1	c_2	c_3	c_4	c_5
–13.688	Sx	0.34333	0.52547	0.39392	0.32289	0.29522
–11.667	Sx	–0.61137	–0.34990	0.02043	0.35739	0.49750
–10.423	Ax	0	0	–0.50289	–0.49709	0
–9.184	Sx	–0.66195	0.31460	0.34908	–0.13435	–0.42784
–1.152	Ax	0	0	–0.49709	0.50289	0
–0.952	Sx	–0.23265	0.57118	–0.24652	–0.29876	0.56531
1.805	Sx	0.12671	–0.41980	0.40223	–0.40088	0.40339

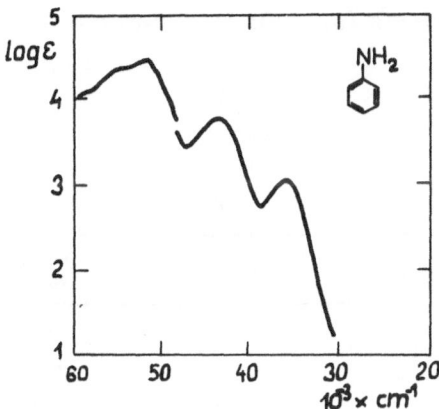

Fig. 3. Electronic spectrum of aniline.

71. Calculate the wavelength of the first intense band in the electronic spectrum of linear polyenes by means of the free electron method (FEMO) and the HMO method. (The HMO orbital energies of the linear polyenes are given by the formula

$$E_j = \alpha + 2 \cos \frac{\pi j}{N+1} \beta,$$

where N is the number of the p_z atomic orbitals and j = 1, 2, ..., N.) Compare and discuss the results obtained by means of the two methods.

72. Calculate the HMO transition moments and the directions of polarization for the $N \rightarrow V_1$ (1 $\rightarrow$ 1') and 1 $\rightarrow$ 2' transitions of pyridine. Consider the pyridine molecule a regular hexagon of side length 1.4 Å. The expansion coefficients are given on page. p. 198.

73. Calculate the HMO transition moment and the direction of polarization for the $N \rightarrow V_1$ transition (1 $\rightarrow$ 1') for γ-thiopyrone. Assume all bonds to be 1.4 Å long.

Expansion coefficients ($\delta_O = 2$; $\delta_S = 0.5$; $\rho_{C=O} = 1.1$; $\rho_{C=S} = 0.9$):

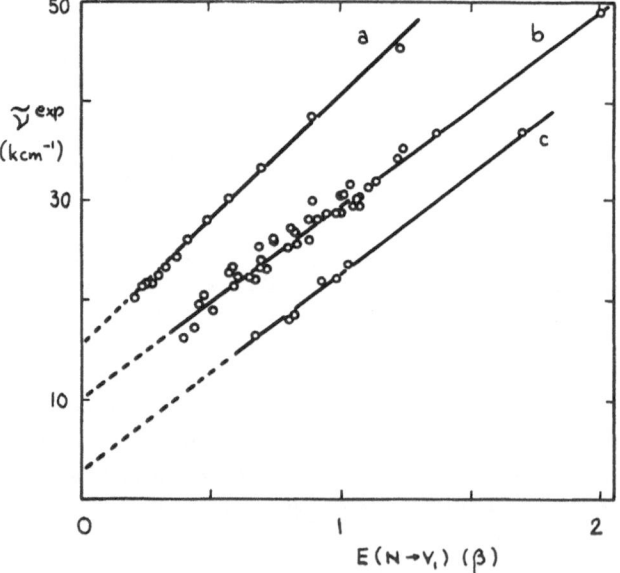

μ	c_μ^1	c_μ^{-1}
1	−0.2485	−0.3796
2	0.1702	0.4807
3	0.3573	0.0399
4	0.0060	−0.5120
5	0.3573	0.0399
6	0.1702	0.4807
7	−0.7906	0.3584

Fig. 4. Wave numbers of maxima of the first intense bands plotted against the N → V$_1$ (HMO) energies. (a) Polyenes, (b) benzenoid hydrocarbons, (c) the tropylium cation and its benzo derivatives.

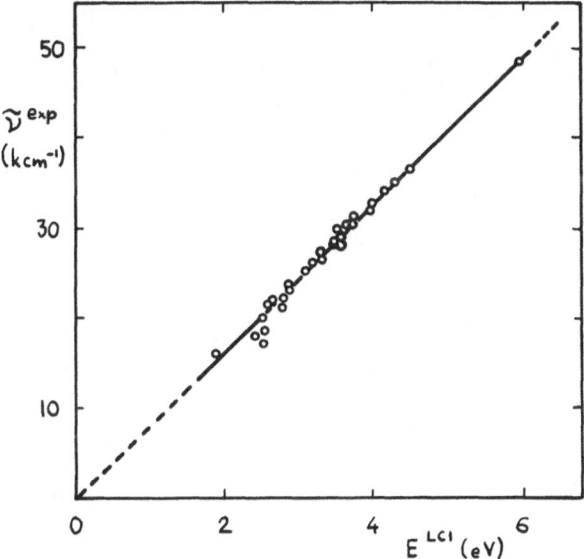

Fig. 5. Wave numbers of maxima of the first intense bands
plotted against LCI-SCF excitation energies.

74. Figures 4 and 5 illustrate the dependence of the measured energies
of the absorption maxima of the first intense bands of benzenoid
hydrocarbons, polyenes, and tropylium compounds on the HMO and
the LCI excitation energies.

In the case of the HMO data, the correlation splits into three dependences according to the class of the substance: (a) polyenes, (b)
benzenoid hydrocarbons, (c) tropylium ions. Correlation of the
LCI excitation energies leads to a single linear dependence for all
these substances. Give a possible reason for this.

4. RADIO-FREQUENCY AND INFRARED SPECTROSCOPIES

75. In the ESR spectra, the following values of the ESR coupling constants were found for a series of radical anions:

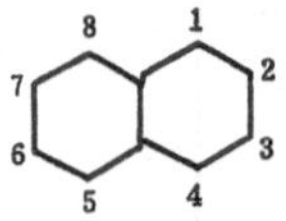

Naphthalene

I

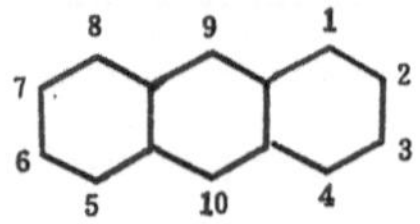

Anthracene

II

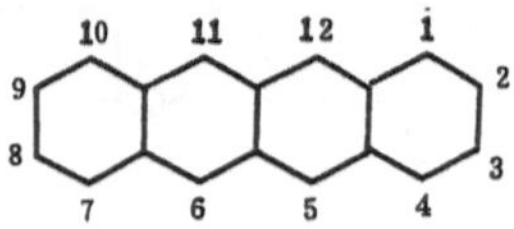

Tetracene

III

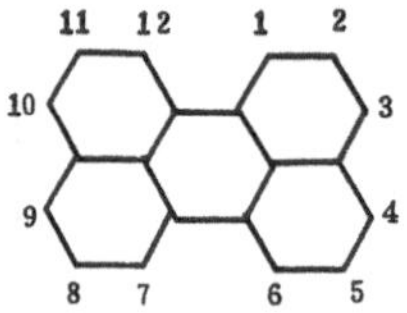

Perylene

IV

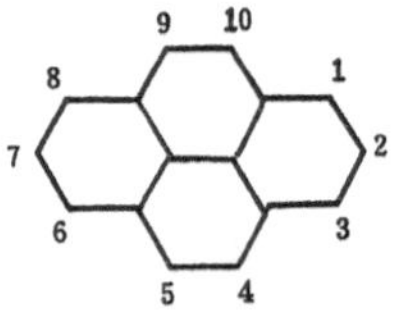

Pyrene

V

Radical anion of substance	Position	a_H (gauss)
I	1	5.01
	2	1.79
II	1	2.74
	2	1.57
	9	5.56
III	1	1.49
	2	1.17
	5	4.25
IV	1	3.09
	2	0.46
	3	3.53
V	1	4.75
	2	1.09
	4	2.08

Suggest   (i) a theoretical characteristic suitable for correlating these data, and

(ii) an empirical equation describing the relationship between theoretical and experimental data.

(According to Ref. 1a)

76. (a) Explain why the ESR spectrum of the naphthalene radical anion consists of 25 lines.

(b) Draw the ESR spectrum when the coupling constants of the protons in the α and β positions have the values of 4.90 and 1.83 gauss, respectively.

(c) Estimate the total extent of the spectrum.

(d) Experimentally it was found that the radical cation has practically the same ESR spectrum as the radical anion. Can you explain this finding by means of the HMO data?

77. Predict the ESR spectrum of the radical anion and the radical cation derived from γ-pyrone. The values of the squares of the expansion coefficients in the positions to which hydrogen is bound are given in the following formulas:

HOMO   LFMO

Within the framework of this example assume that the squares of the HMO expansion coefficients are a measure of the spin densities, although we know that this is not the case when the expansion coefficients become very small or equal to zero. Assume that the constant Q in the relation between the ESR coupling constant a_H and the spin density ρ has the value of 30 gauss ($a_H = Q\rho$).

78. The following ESR coupling constants were measured for the allyl radical[13] (in gauss): $a_1 = 13.936$, $a_1' = 14.836$, and $a_2 = 4.066$. a_2 concerns the hydrogen attached to the central carbon; owing to the

bent form of the radical the remaining four hydrogens are not equivalent but exhibit two coupling constants a_1 and a_1'.

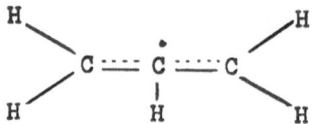

Attempt to substantiate this finding (a) within the framework of the
HMO method (for expansion coefficients, see p. 191), and (b) by
application of the McLachlan equation (use the value $\lambda = 1.06$; for
expansion coefficients, see p. 191).

Atom – atom polarizabilities for the allyl system:

	1	2	3
1	0.4419		
2	−0.1768	0.3536	
3	−0.2632	−0.1768	0.4419

79. The following values of the NQR frequencies were determined experimentally for a series of chloro derivatives[14] (Mc/sec):

2-chloropyridine	34.17	2-chloroquinoline	33.29
3-chloropyridine	35.24	6-chloroquinoline	34.60
4-chloropyridine	34.89	7-chloroquinoline	34.68

Investigate whether these values are correlated with the density
of the π-electrons on the Cl atom.

π-electron densities on the Cl atom (HMO parameters employed:
$\delta_{Cl} = 2$; $\delta_N = 0.5$; $\rho_{CCl} = 0.4$):

2-chloropyridine	1.9827	2-chloroquinoline	1.9815
3-chloropyridine	1.9849	6-chloroquinoline	1.9847
4-chloropyridine	1.9831	7-chloroquinoline	1.9841

80. A correlation has been established[15] (cf. also Ref. 1a) between the
wave numbers of the stretching vibrations of the C=O bonds of
carbonyl compounds and the corresponding bond orders $p_{C=O}$ (Fig. 6).
(The bond orders were calculated for the following parameters: $\alpha_O = \alpha + 1.3\beta$, $\beta_{CO} = \sqrt{2}\beta$, $\alpha_{C(O)} = \alpha + 0.2\beta$.)

Ascertain the agreement between the theoretical and the experi-

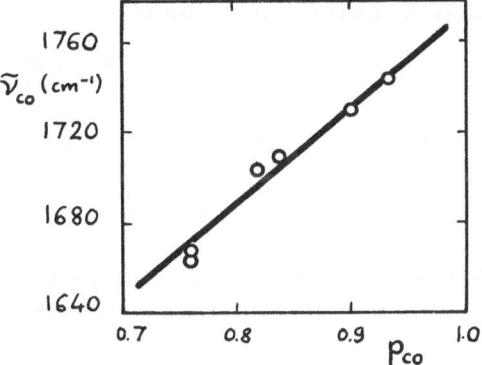

Fig. 6. Wave numbers of the carbonyl stretching vibrations plotted against the $C=O$ bond
orders.

mental (1670 cm^{-1}) values for 9–benzoylanthracene. (The bond
order $p_{C=O}$ is equal to 0.727.)

5. CHEMICAL REACTIVITY

81. What is the probable composition and structure of the substance
formed by the reaction of pyridine with (a) a solution of sodium
amide in liquid ammonia, (b) gaseous bromine at 350° C, and (c)
a nitration mixture. In case (c) take into consideration that either
the free base or the protonated form reacts. For the prediction
employ the data of the molecular diagrams and of localization energies given below.

Atom localization energies (in β-units):

	A_e	A_r	A_n
2	2.672	2.512	2.353
3	2.538	2.538	2.538
4	2.701	2.537	2.374

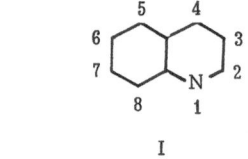

The atom localization energies of the protonated form are qualitatively the same for the individual positions as for the free base.

82. Nitration of quinoline (I) with a solution of nitric acid in glacial
acetic yields predominantly two isomers; the 5- and 8-nitro derivatives. Using the atom localization energies (A_e, given in paren-
theses) and the π-electron densities ascertain whether the theoretical reactivity indices are in agreement with the experimentally
established course of the reaction. (The molecular diagram for
the protonated form is qualitatively very similar.)

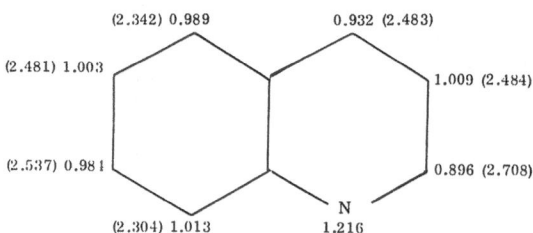

I

83. Predict the sequence of the reactivity values of the individual positions of picene for electrophilic, nucleophilic, and radical substitution. Which is the probable preferential course of the reaction
with OsO_4 and maleic anhydride? Ortho- and para-localization
energies are given in the lower diagram (β-units).

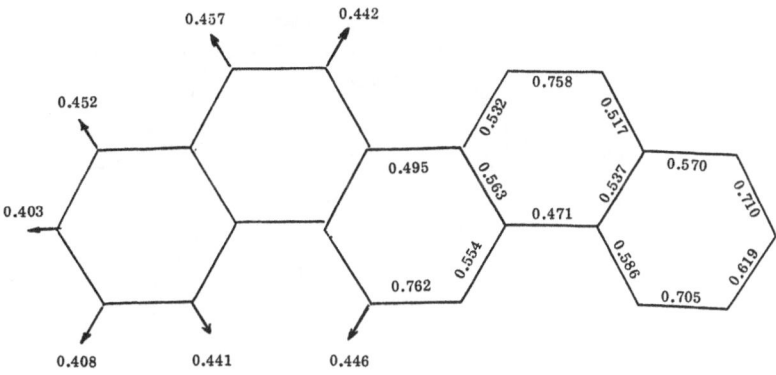

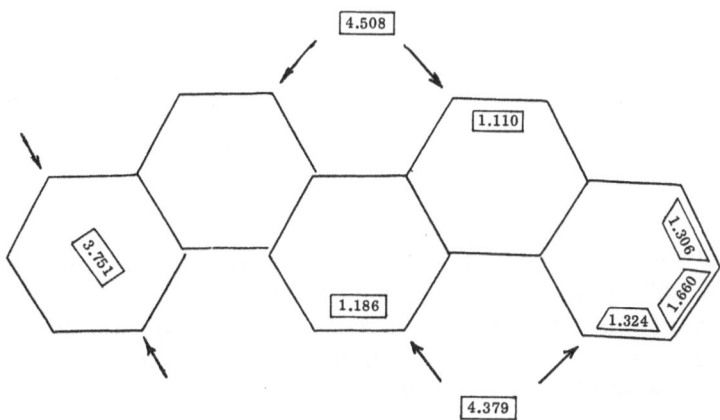

84. Predict the position of the probable maximum reactivity in elec-
    trophilic, nucleophilic, and radical substitution on the basis of the
    molecular diagram and the localization energies of azulene. It was
    found experimentally that positions 1 and 3 are centers of electro-
    philic reactions (nitration), and position 4 and 8 centers of nucleo-
    philic reactions (amidation). Is the theory in agreement with the
    experiment?

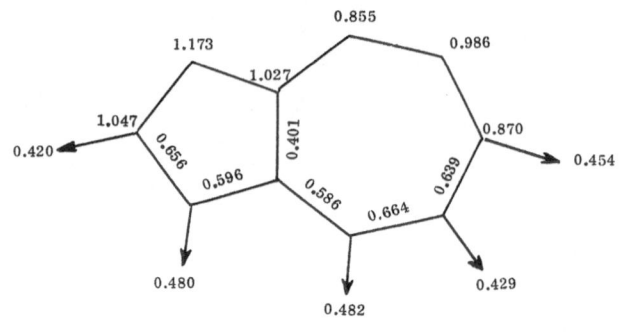

	A_e	A_r	A_n
1	1.924	2.262	2.600
2	2.346	2.346	2.346
4	2.551	2.240	1.929
5	2.341	2.341	2.341
6	2.730	2.359	1.988

85. Molecular diagram of the trithione derived from 1,3-dithiylium (the calculation was performed for the following values of the HMO parameters: $\alpha_{-S-} = \alpha + \beta$, $\alpha_{=S} = \alpha + 0.5\beta$, $\beta_{C-S} = 0.6\beta$, $\beta_{C=S} = 0.9\beta$, i.e., for a model in which the participation of the d orbital is not explicitly assumed):

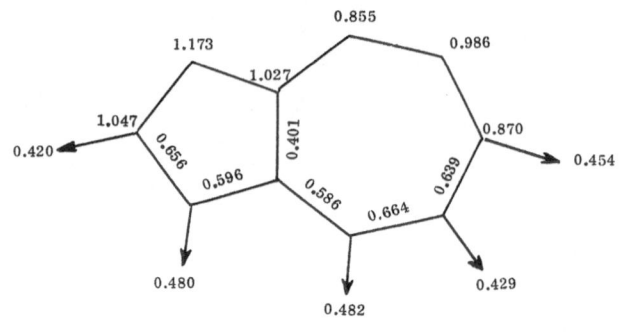

(a) What are the conditions for nucleophilic, electrophilic, and radical substitution reactions of this substance?

(b) Which reaction is likely to set in on addition of bromine water to this substance?

86. It was found that, unlike solutions of 6-azauracil, aqueous solutions of 5-azauracil are unstable. Try to explain this behavior by means of the molecular diagrams.

87. In which positions of the hydrocarbons I-III will the Diels – Alder reaction with maleic anhydride proceed preferentially ?

π-electron bonding energies (in β units):

Benzene	8.000	Phenanthrene	19.449
Ethylene	2.000	Pyrene	22.506
Naphthalene	13.683	Benzanthracene	25.101

Tetracene	24.931
I	30.879
II	30.726
III	28.222

88. Ozone splits the C = C double bonds, and carbonyl compounds are formed. The primary step of the reaction appears to be the formation of the π-complex

Using the ortho-localization energies, predict the bonds at which the following hydrocarbons will probably be attacked (the bond orders are indicated along the bonds inside the formulas; the ortho-localization energies are given outside the formulas and are underlined):

Find out at which bonds the above-mentioned reaction is likely to proceed preferentially if you relate your prediction to the values of the bond orders. Compare the two predictions. Experimentally it was found that the oxidation sets in at the following bonds:

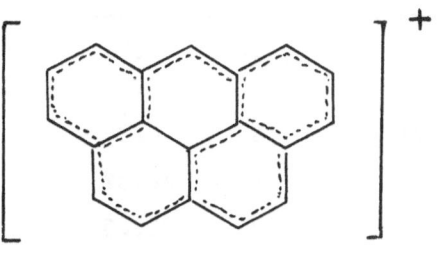

89. The chlorine in 9-chlorophenanthrene cannot be substituted nucleophilically. However, if nitrogen is introduced into the rings as a heteroatom, chlorine can be replaced by the $OH^{(-)}$ ion in a nucleophilic reaction. Decide by means of the Longuet-Higgins NBMO theory[16] in which position the introduction of the heteroatom will enhance the nucleophilic substitution to the greatest extent.[17] (See also Exercise 52.)

90. Predict the center of the nucleophilic reactivity in the cation I; for the calculation employ a procedure not requiring the solution of the secular equations.

I

91. What will probably be the structure of the product of the coupling reaction of 2-hydroxydiphenylene with benzenediazonium chloride? This reaction is in principle an electrophilic substitution. Do not employ the molecular diagram to characterize the reactive centers

of the 2-hydroxydiphenylene molecule; base your prediction upon the NBMO method of Longuet-Higgins.[17]

It is your task to complete the following reaction scheme:

92. In table the logarithms of rate constants (log k) of electrophilic bromination of benzenoid hydrocarbons (according to Ref. 18) are summarized, together with HMO and SCF atomic localization energies for electrophilic substitution. Use these data and

(i)   verify whether there exists a correlation between log k and A^{HMO} and A^{SCF},

(ii) decide whether the difference between the plots in (i) is due to the steric hindrance or to the neglect of the electronic repulsion term in the HMO method,

(iii) predict the rate constant of bromination for position 1 in anthracene. Base this prediction on both the HMO and SCF localization energies.

Electrophilic bromination of benzenoid hydrocarbons

Compound	Position	log k^a (Ref.18)	Localization energy	
			HMO (β-units)	SCF (eV)
Naphthalene	1	5.26	2.299	24.643
	2	3.27	2.480	25.199
Phenanthrene	9	6.35	2.298	24.497
Anthracene	9	12.37	2.013	23.502
Pyrene	1	10.63	2.190	23.855
Chrysene	6	7.57	2.254	—
1,2-Benzanthracene	7	11.16	2.049	23.49
Diphenyl	4	3.64	2.447	24.940
Anthracene	1	?	2.231	24.213

^{a}Relative values, for benzene log k = 1.

93. Employ the frontier orbital theory [K. Fukui, Fort. Chem. Forsch. 15:1 (1970)] and give your arguments (a) for the determination which of the following two structures of the charge-transfer complex of benzene with the Ag^+ ion is more probable (the electronic configuration of Ag is . . . $4s^2 4p^6 4d^{10} 5s$)

(I)   (II)

and (b) why in contrast to the 7-norbornenyl anion (I), the 7-norbornenyl cation (II) is stabilized by the interaction between the position 7 and the double bond:

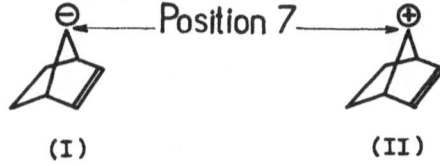

(I)   (II)

94. Based on the squares of the extended Hückel expansion coefficients of the lowest free molecular orbital in 2-amyl chloride (I) and t-amyl chloride (II), give the theoretical foundation of the Saytzeff rule (preferential hydrogen abstraction from the carbon atom carrying the maximum number of alkyl substituents). The values of squares of expansion coefficients for the possible sites of attack are underlined [K. Fukui, H. Hao, and H. Fujimoto, Bull. Chem. Soc. Japan 42:348 (1969)].

I

.0001
H

Cl
.0291   .0141
H   H

CH₃

.0003   .0909   .0003   .0928
H H   H H

H .0002

tert.

II

95. For a series of 1,2- and 2,3-bridged cyclopropane derivatives, the following relative rate constants of solvolysis were determined:

C_3H_7

HOH$_2$C   CH$_2$OTs   CH$_2$OTs

H C_3H_7 H   H

k_{rel}
(HOAc)

$\frac{1}{80}$ $\frac{1}{40}$

CH$_2$OTs   CH$_2$OTs

H   H

k_{rel}
(HOAc)

1   5

CH$_2$OPNB   CH$_2$OPNB

k_{rel}
(60% aq. acetone)   550   1

(Ts stands for the p-toluensulfonyl and PNB for p-nitrobenzoyl group)

Suggest a simple way how these data could be interpreted by means of MO calculations.

Hint: Compare the bond populations in methylcyclopropane and the cyclopropylmethyl cation [K. B. Wiberg, Tetrahedron 24:1083 (1968)]:

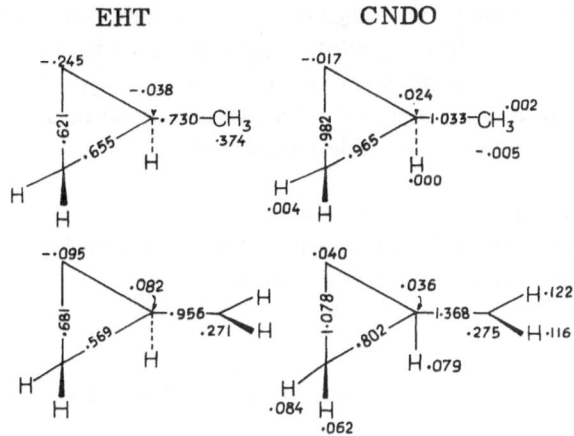

Charge densities are given at atoms, overlap population
(EHT) and bond indices (CNDO) are given in bonds

96. The following figure presents the correlation diagram based on the
extended Hückel calculations for N_2H_2 orbital energies in planar

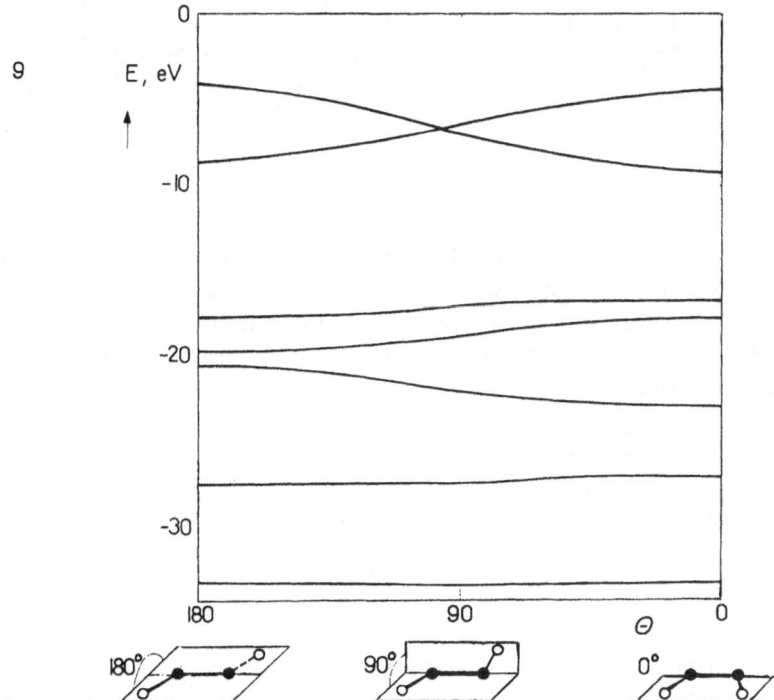

(cis and trans) and nonplanar geometries (B. M. Gimarc, J. Am. Chem. Soc. 92:266 (1970)]. By means of this diagram determine whether the concerted cis – trans (C_{2v} – C_{2h}) isomerization of N_2H_2 via a nonplanar (C_2) transition state is a symmetry-allowed or a symmetry-forbidden reaction.

Assuming that the correlation diagram for H_2O_2 is qualitatively similar to that in the figure above, also make the prediction for the cis – trans isomerization of H_2O_2.

6. MISCELLANEOUS

97. Estimate the oxidation – reduction potentials of the quinones I-III. Since we have here a quantity characterizing the equilibrium state, it would be appropriate to base the estimate upon the differences of the π -electron energies of the oxidized and of the reduced form.

I   II   III

It was found, however, that these differences are linearly dependent on the bicentric localization energies of the parent hydrocarbons for the positions where oxygen atoms are bound in the qui - nones. In the case of the quinones investigated we have to do with ortho- and para-localization energies; the corresponding torsos (Ia, IIa, IIIa) are

Ia   IIa   IIIa

The relation between the oxidation – reduction potentials and the respective bicentric localization energies is governed by the following empiric equations: [19]

ortho-Quinones: $E(V) = 0.617 (A_o + 2) - 1.666$

para-Quinones: $E(V) = 0.598 A_p - 1.687$

The π-electron energies W are:

$W = 20 \alpha + 28.222 \beta$

$W = 22 \alpha + 30.942 \beta$

$W = 22 \alpha + 30.544 \beta$

$W = 18 \alpha + 25.192 \beta$

$W = 6 \alpha + 8 \beta$

$W = 14 \alpha + 19.448 \beta$

$W = 10 \alpha + 13.683 \beta$

$W = 2 \alpha + 2.000 \beta$

98. It has been found that the polarographic half-wave potentials of the cathodic waves ($E_{1/2}$, V) of polynuclear benzenoid hydrocarbons (0.18 M (C_4H_9)$_4$ NI, 96% dioxane)[20] are linearly dependent on the energy of the lowest unoccupied π-molecular orbital (k_{-1}):

$$E_{1/2} = 2.72 k_{-1} - 0.79$$

(correlation coefficient r = 0.962; number of substances n = 14)

(a) Predict the value of $E_{1/2}$ for naphthalene and anthracene. The orbital energies are given on p. 189.

(b)   Assume that a difference of 0.05 V in the half-wave potential
      in needed to distinguish between waves; is a successful dis-
      tinction between tetracene and pentacene probable?

(c)   Watson and Matsen[21] found that for benzenoid hydrocarbons the
      polarographic half-wave potentials are linearly dependent on
      the excitation energies of p bands. Attempt to explain this re-
      sult in terms of HMO theory. Can we expect the $E_{1/2}$ values
      of benzenoid hydrocarbons also to be correlated with ioniza-
      tion potentials and with the excitation energies of the first ab-
      sorption bands of the charge-transfer complexes? What form
      of the dependence is to be expected?

(d)   Which of the dependences mentioned in (c) will be satisfied by
      nonalternant hydrocarbons?

99. Calculate the π-electron contribution to the dipole moment (a) for
     the ground state of pyridine and (b) for the singly excited $1 \rightarrow 1'$
     state.

     The HMO π-electron densities ($\delta_N = 0.5$) are:

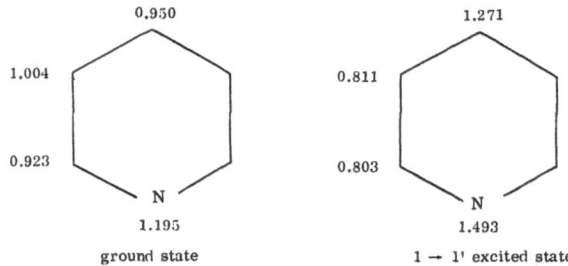

As far as the geometry of pyridine is concerned, assume a regular
hexagon of side 1.4 Å. Multiply the result by the factor 4.80 to
obtain the dipole moment in D units.

100. By means of the bond orders (see p. 204), predict the length of the
      $C - C$ bonds $1-2$, $2-3$, and $11-12$ for anthracene. Compare the
      values obtained in this way with the experimental data:[22]

Bond   1-2 1.37_5 Å

Bond   2-3   1.41 Å

Bond   11-12   1.44 Å

101. Decide by means of Hückel's (4n + 2) rule which of the following
     hydrocarbons is aromatic (A). Indicate especially those hydro-
     carbons to which (with regard to the definition) this rule cannot
     be applied.

(a)    (b)    (c)    (d)

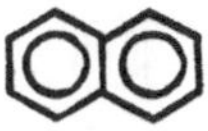

(e)    (f)    (g)    (h)

(i)    (j)    (k)   (1)

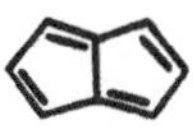

(m)    (n) 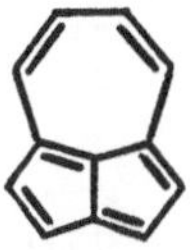   (o)    (p)    (q)

(r)    (s) 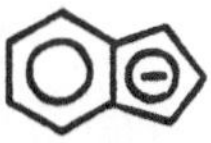   (t)    (u)

102. Decide by means of Craig's rule which of the following hydrocar-
     bons is aromatic. Indicate especially those hydrocarbons to which
     this rule is not applicable.

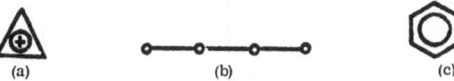

(a)   (b)   (c)

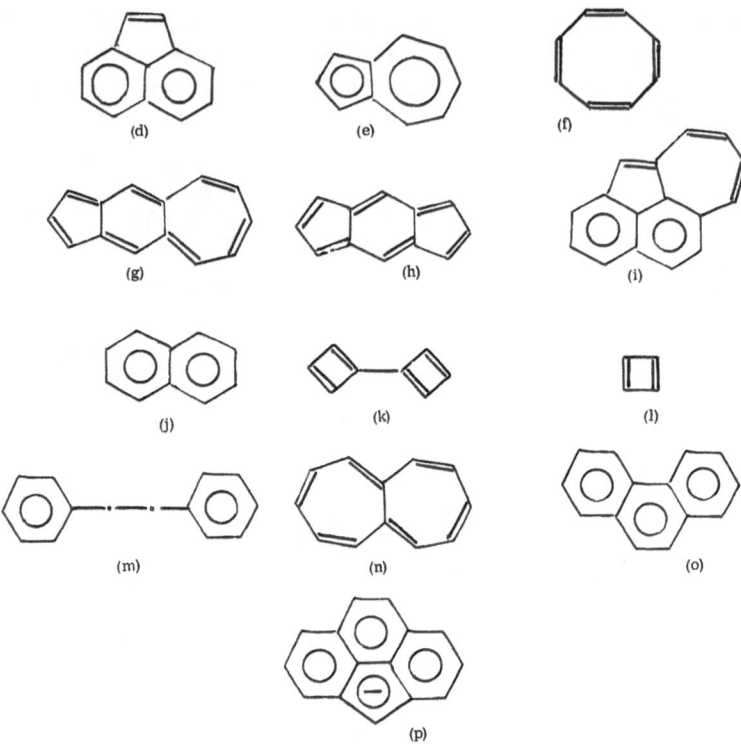

103. Attempts to synthesize the cation I have failed. The HMO calculation (Longuet-Higgins model for the AO of sulfur, δ_{CS} = 0.6) re-
sulted in the following values of the orbital energies:

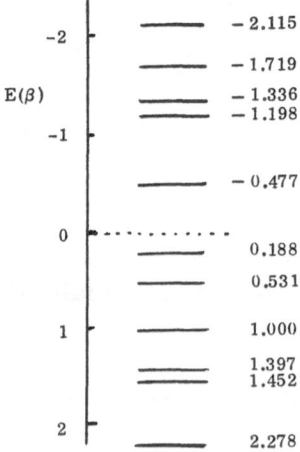

What is the probable cause of the instability of the cation? Do you
think that (from the viewpoint of the given values of the orbital
energies) a synthesis of the anion II is more promising?

I   II

04. Predict on the basis of Pullman's criteria[23] whether the following
hydrocarbons are carcinogenic or noncarcinogenic. The values of
the ortho-localization energies of the K-spheres and of the para-
localization energies of the L-spheres (in β units) are directly
inscribed in the formulas; the values of the atom localization energies are boxed:

II. Notes on the Solutions and Results

1. (a)

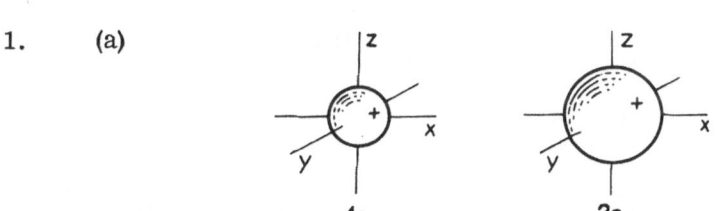

1s   2s

(b)

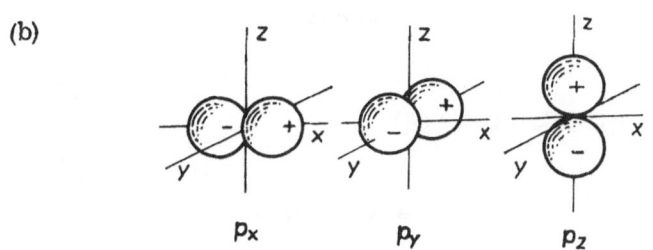

p_x p_y p_z

(c)

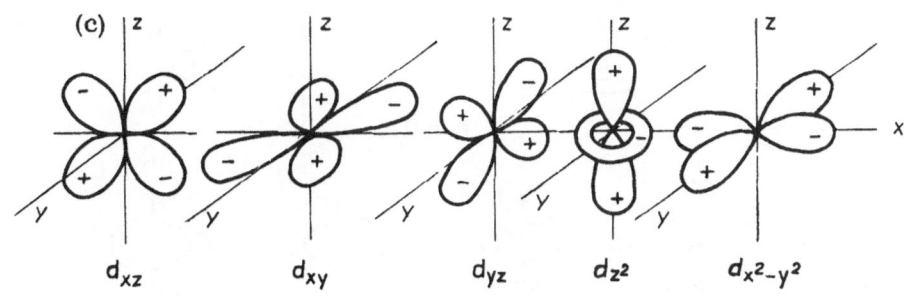

d_{xz} d_{xy} d_{yz} d_{z^2} $d_{x^2-y^2}$

2.

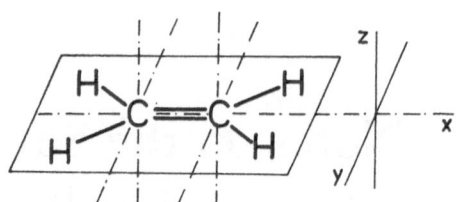

View in the direction of the   View in the direction of the
y-axis   z-axis

H – C = C – H

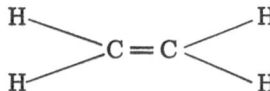

(a)

bonding

   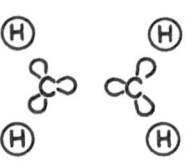

antibonding

(b)

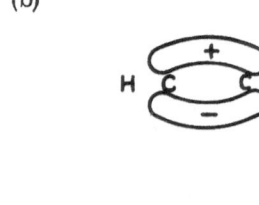   

bonding

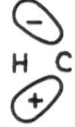   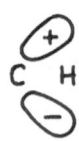   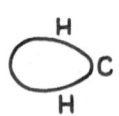   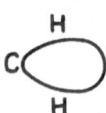

antibonding

3.

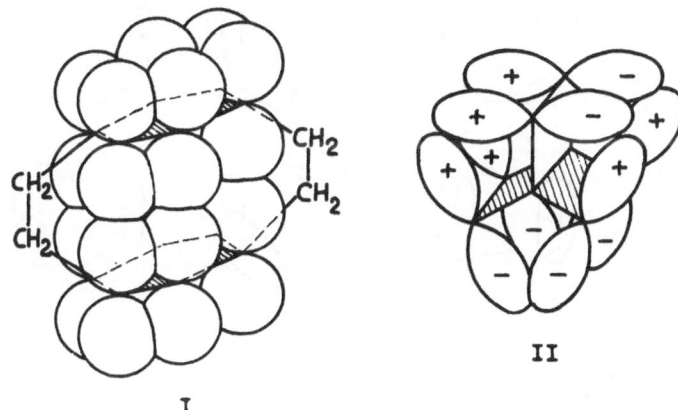

I   II

Note that there are six overlaps in system II which could be labeled as nonclassical π-bonds, and that two of them are antibonding. (II represents one of two degenerated arrangements.)

4.   (a)

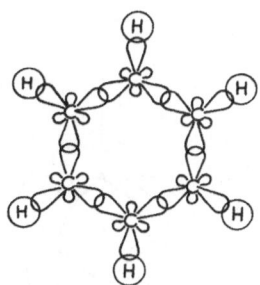

        (b)

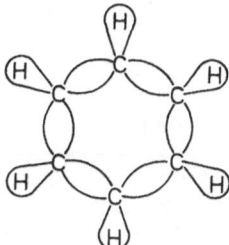

(c)

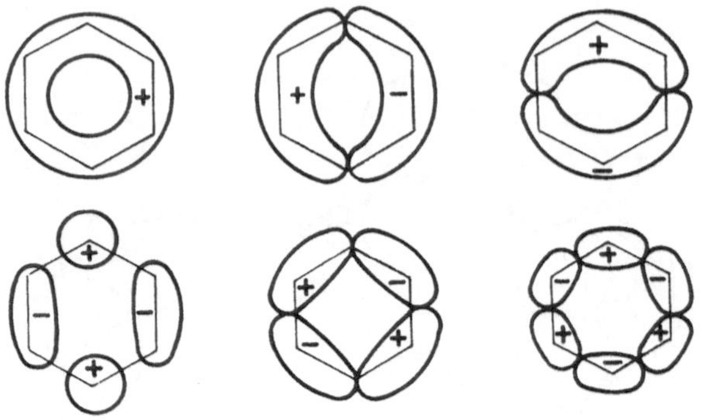

5.

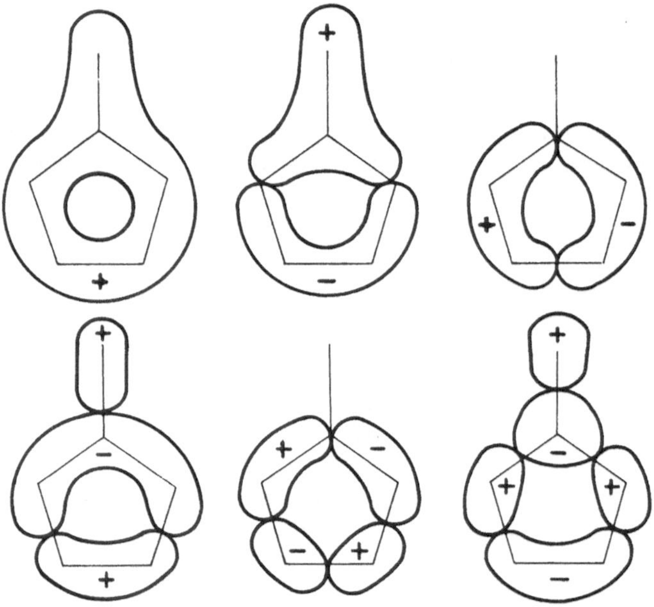

6.   (a)

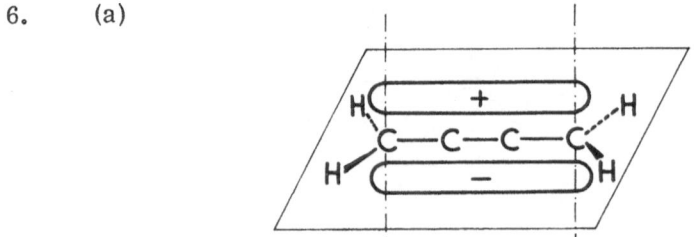

(b)

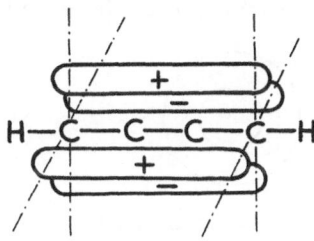

(c)

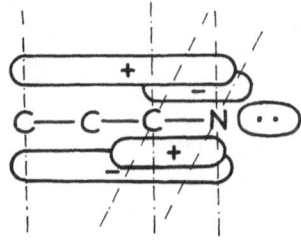

(d)

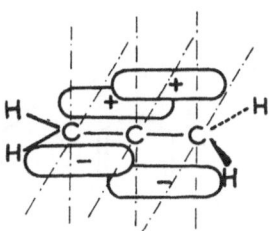

7.   (a) View in the direction of the y-axis (the coordinate system
          is the same as in Exercises 1 and 2)

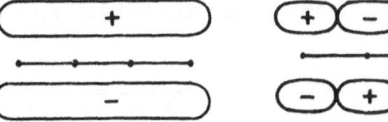

     (b) View in the direction of the y-axis

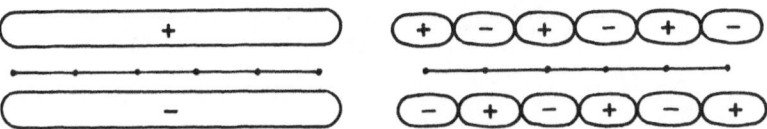

(c) View in the direction of the z-axis

(d) View in the direction of the z-axis

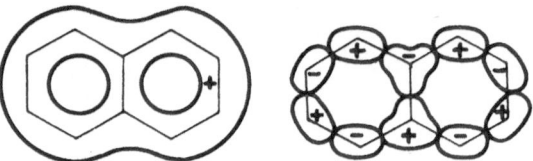

No nodal plane with the energetically most favorable π-MO; maximum possible number of nodal planes with the energetically least favorable π-MO.

8.

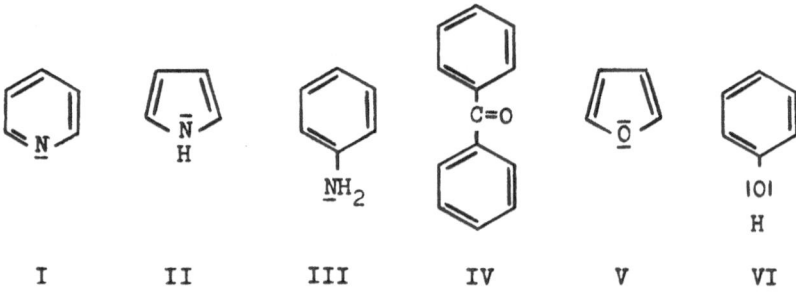

It is assumed that in these compounds the heteroatom is in the sp^2 hybrid state (the $2p_z$ atomic and the hybrid sp^2 orbitals are indicated):

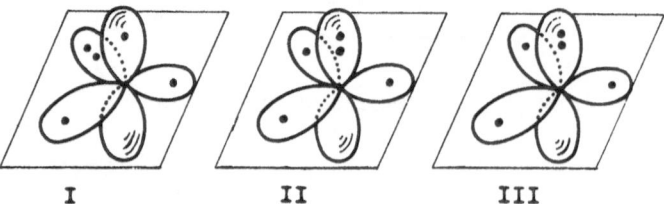

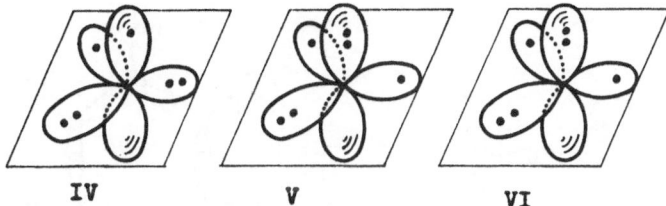

IV

V

VI

View in the direction of the z-axis.

(1a) Overlap of atomic orbitals which form σ-bonds (sp^2 hybrid orbitals):

I

II

III

IV

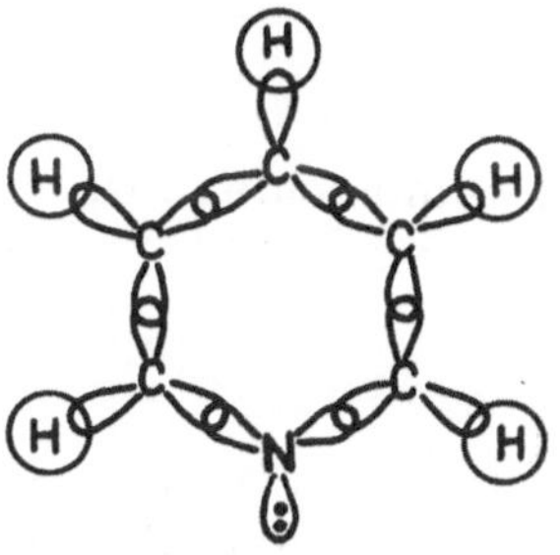

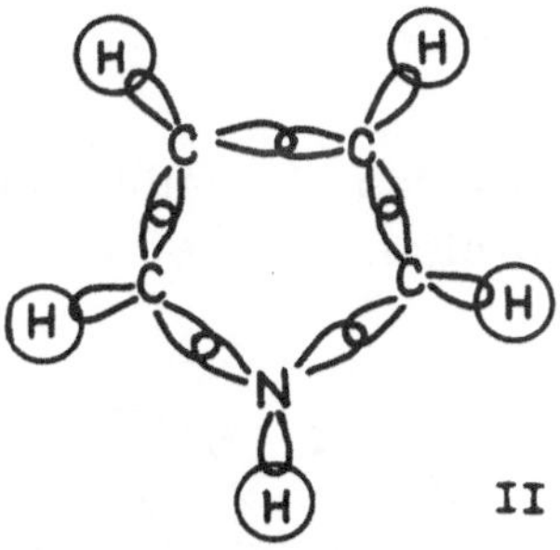

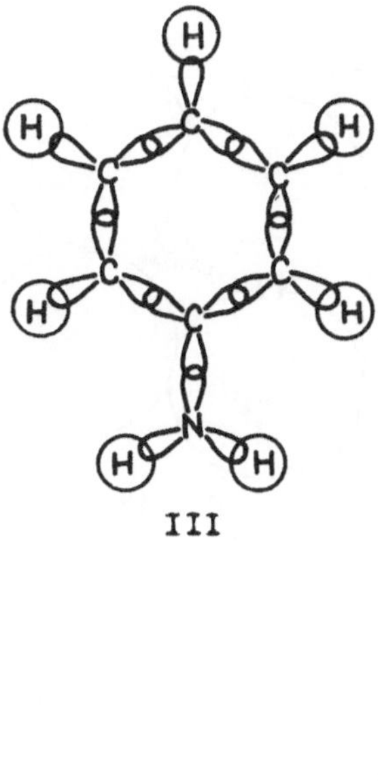

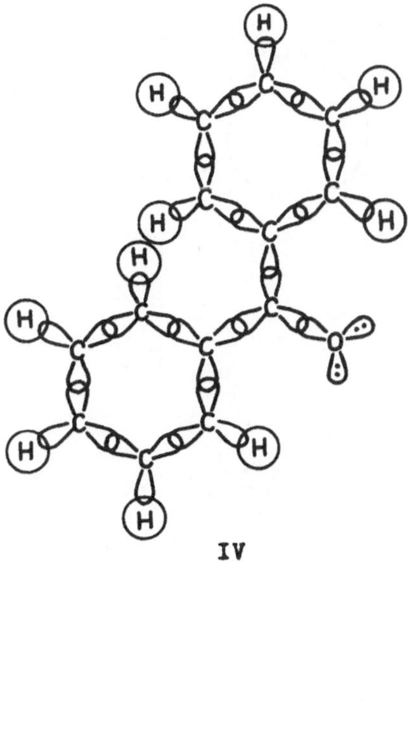

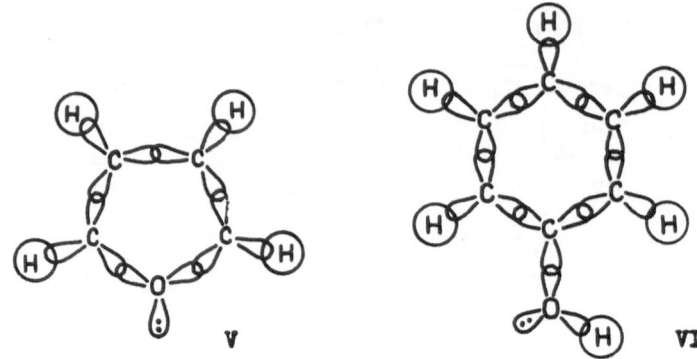

(1b) Overlap of atomic orbitals forming π-molecular orbitals ($2p_z$ orbitals):

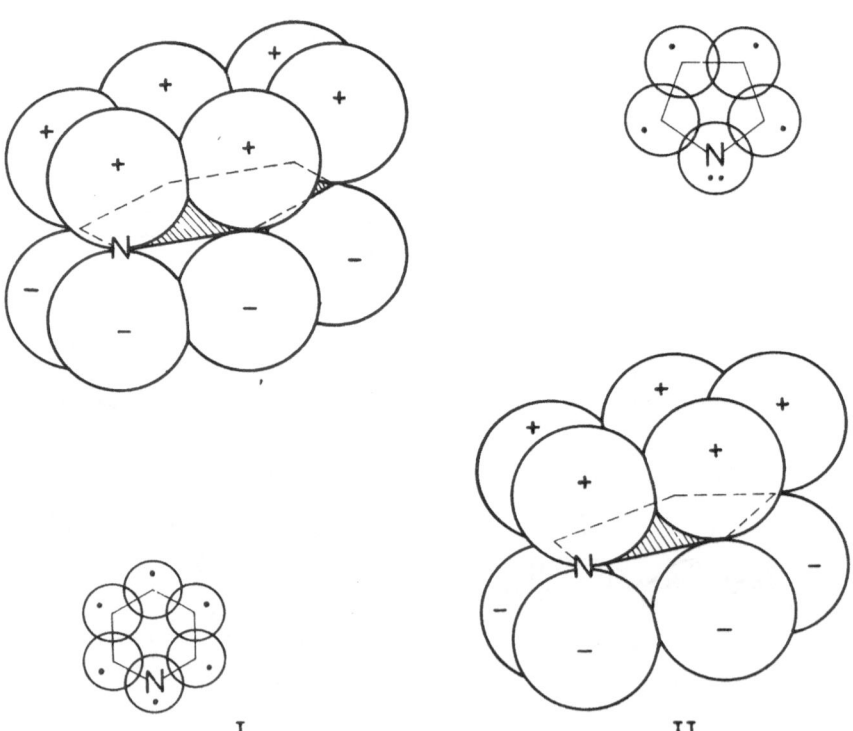

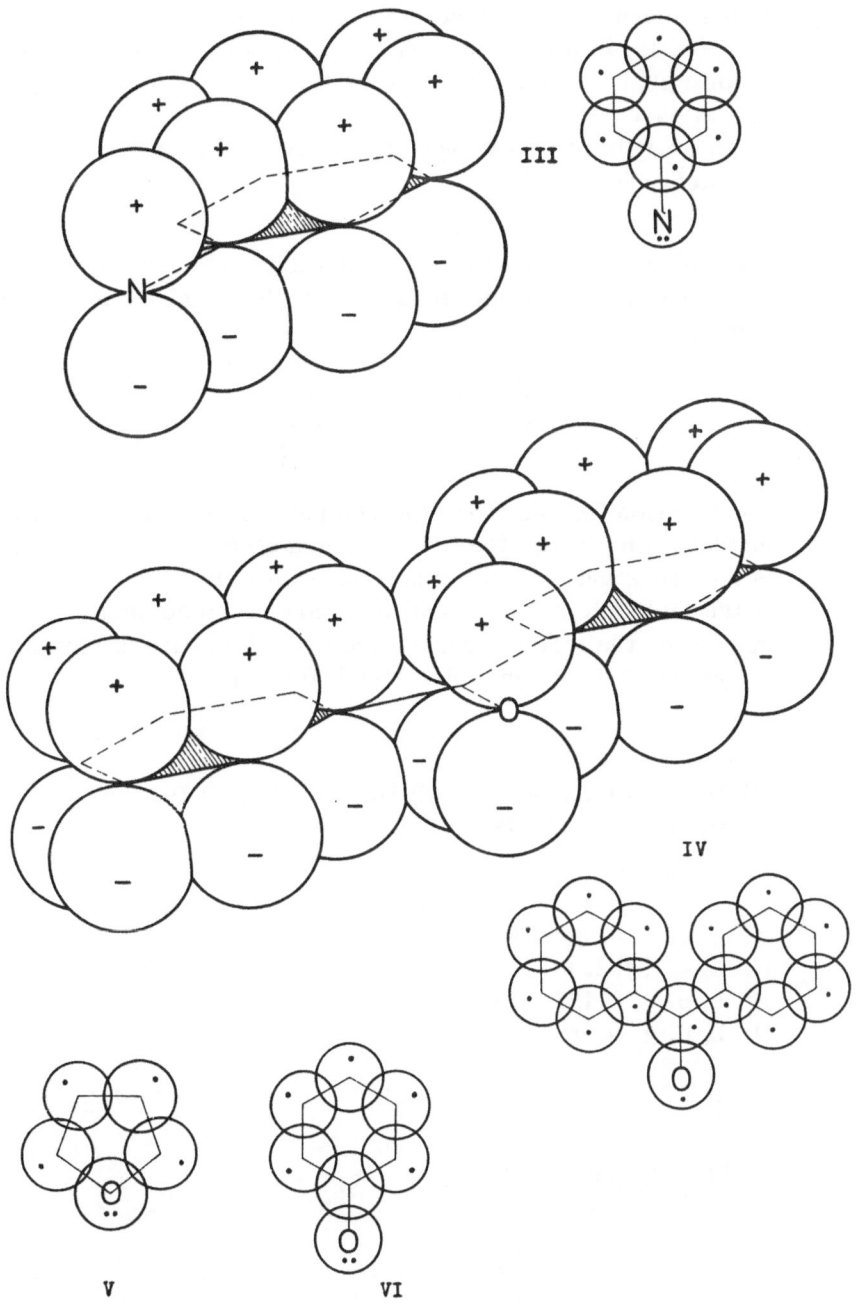

III

IV

V VI

(2) In benzophenone both lone pairs are sp² hybrid nonbonding atomic orbitals, in phenol one pair is in an sp² hybrid nonbonding
atomic orbital whereas the other takes part in the conjugation
and is part of the π-molecular orbitals (see diagrams). (This
is an oversimplified view; for a more realistic description,
see Ref. 11).

9.   The different nature of the nitrogen lone electron pairs in pyridine and pyrrol cannot be seen from the usual structural formulas:

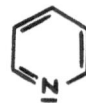   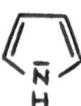

The notation for the free electron pair is the same in both cases.
In reality, however, the free nitrogen electron pair in pyridine
is a nonbonding sp² hybrid atomic orbital which lies in the plane
of the molecule but does not take part in the conjugation. In
pyrrol, on the other hand, the free electron pair is part of the
π-system and the symbol — stands for a $2p_z$ AO.

10.   (a) D_{6h}, (b) C_{2v}, (c) C_{2h}, (d) D_{3h}, (e) C_{2v}, (f) D_{3h}, (g) C_{1h},
(h) D_{10h}, (i) C_{2v}, (j) D_{2h}, (k) D_{2h}.

11.   A: a, b, c, d, g, h, k, m
N: e, f, i, j, l, n
E: a, c, e, f, i, j, k, l, n
O: b, d, g, h, m

12.
$$\int \varphi_1 \varphi_2 \, d\tau = \int \left[\tfrac{1}{2} (\chi_s + \chi_{p_x} + \chi_{p_y} + \chi_{p_z}) \right] \times$$

$$\times \left[\tfrac{1}{2} (\chi_s - \chi_{p_x} - \chi_{p_y} + \chi_{p_z}) \right] d\tau =$$

$$= \tfrac{1}{4} \int (\chi_s)^2 \, d\tau - \tfrac{1}{4} \int \chi_s \chi_{p_x} \, d\tau -$$

$$- \tfrac{1}{4} \int \chi_s \chi_{p_y} \, d\tau + \tfrac{1}{4} \int \chi_s \chi_{p_z} \, d\tau +$$

$$+ \tfrac{1}{4} \int \chi_{p_x} \chi_s \, d\tau + \ldots =$$

$$= \tfrac{1}{4} - \tfrac{1}{4} - \tfrac{1}{4} + \tfrac{1}{4} = 0$$

Similarly for the remaining sp^3 orbitals.

13.   (a)

$$-kc_1 + c_2 = 0$$
$$-kc_2 + c_1 + c_3 = 0$$
$$-kc_3 + c_2 + c_4 = 0$$
$$-kc_4 + c_3 = 0$$

$$\begin{vmatrix} -k & 1 & 0 & 0 \\ 1 & -k & 1 & 0 \\ 0 & 1 & -k & 1 \\ 0 & 0 & 1 & -k \end{vmatrix} \qquad \begin{pmatrix} 0 & 1 & 0 & 0 \\ 1 & 0 & 1 & 0 \\ 0 & 1 & 0 & 1 \\ 0 & 0 & 1 & 0 \end{pmatrix}$$

(b)

$$-kc_1 + c_2 = 0$$
$$-kc_2 + c_1 + c_3 + c_7 = 0$$
$$-kc_3 + c_2 + c_4 = 0$$
$$-kc_4 + c_3 + c_5 = 0$$
$$-kc_5 + c_4 + c_6 = 0$$
$$-kc_6 + c_5 + c_7 = 0$$
$$-kc_7 + c_2 + c_6 + c_8 = 0$$
$$-kc_8 + c_7 = 0$$

$$\begin{vmatrix} -k & 1 & 0 & 0 & 0 & 0 & 0 & 0 \\ 1 & -k & 1 & 0 & 0 & 0 & 1 & 0 \\ 0 & 1 & -k & 1 & 0 & 0 & 0 & 0 \\ 0 & 0 & 1 & -k & 1 & 0 & 0 & 0 \\ 0 & 0 & 0 & 1 & -k & 1 & 0 & 0 \\ 0 & 0 & 0 & 0 & 1 & -k & 1 & 0 \\ 0 & 1 & 0 & 0 & 0 & 1 & -k & 1 \\ 0 & 0 & 0 & 0 & 0 & 0 & 1 & -k \end{vmatrix}$$

$$\begin{pmatrix} 0 & 1 & 0 & 0 & 0 & 0 & 0 & 0 \\ 1 & 0 & 1 & 0 & 0 & 0 & 1 & 0 \\ 0 & 1 & 0 & 1 & 0 & 0 & 0 & 0 \\ 0 & 0 & 1 & 0 & 1 & 0 & 0 & 0 \\ 0 & 0 & 0 & 1 & 0 & 1 & 0 & 0 \\ 0 & 0 & 0 & 0 & 1 & 0 & 1 & 0 \\ 0 & 1 & 0 & 0 & 0 & 1 & 0 & 1 \\ 0 & 0 & 0 & 0 & 0 & 0 & 1 & 0 \end{pmatrix}$$

(c)

$$-kc_1 + c_2 + c_{10} = 0$$
$$-kc_2 + c_1 + c_3 = 0$$
$$-kc_3 + c_2 + c_4 = 0$$
$$-kc_4 + c_3 + c_5 = 0$$
$$-kc_5 + c_4 + c_6 + c_{10} = 0$$
$$-kc_6 + c_5 + c_7 = 0$$
$$-kc_7 + c_6 + c_8 = 0$$
$$-kc_8 + c_7 + c_9 = 0$$
$$-kc_9 + c_8 + c_{10} = 0$$
$$-kc_{10} + c_1 + c_5 + c_9 = 0$$

$$
\begin{vmatrix}
-k & 1 & 0 & 0 & 0 & 0 & 0 & 0 & 0 & 1 \\
1 & -k & 1 & 0 & 0 & 0 & 0 & 0 & 0 & 0 \\
0 & 1 & -k & 1 & 0 & 0 & 0 & 0 & 0 & 0 \\
0 & 0 & 1 & -k & 1 & 0 & 0 & 0 & 0 & 0 \\
0 & 0 & 0 & 1 & -k & 1 & 0 & 0 & 0 & 1 \\
0 & 0 & 0 & 0 & 1 & -k & 1 & 0 & 0 & 0 \\
0 & 0 & 0 & 0 & 0 & 1 & -k & 1 & 0 & 0 \\
0 & 0 & 0 & 0 & 0 & 0 & 1 & -k & 1 & 0 \\
0 & 0 & 0 & 0 & 0 & 0 & 0 & 1 & -k & 1 \\
1 & 0 & 0 & 0 & 1 & 0 & 0 & 0 & 1 & -k
\end{vmatrix}
$$

$$
\begin{pmatrix}
0 & 1 & 0 & 0 & 0 & 0 & 0 & 0 & 0 & 1 \\
1 & 0 & 1 & 0 & 0 & 0 & 0 & 0 & 0 & 0 \\
0 & 1 & 0 & 1 & 0 & 0 & 0 & 0 & 0 & 0 \\
0 & 0 & 1 & 0 & 1 & 0 & 0 & 0 & 0 & 0 \\
0 & 0 & 0 & 1 & 0 & 1 & 0 & 0 & 0 & 1 \\
0 & 0 & 0 & 0 & 1 & 0 & 1 & 0 & 0 & 0 \\
0 & 0 & 0 & 0 & 0 & 1 & 0 & 1 & 0 & 0 \\
0 & 0 & 0 & 0 & 0 & 0 & 1 & 0 & 1 & 0 \\
0 & 0 & 0 & 0 & 0 & 0 & 0 & 1 & 0 & 1 \\
1 & 0 & 0 & 0 & 1 & 0 & 0 & 0 & 1 & 0
\end{pmatrix}
$$

1. Using the σ_x – plane of symmetry:

$$S_x: \quad c_1 = c_4; \; c_2 = c_3; \; c_8 = c_7; \; c_9 = c_6; \; c_{10} = c_5$$

$$-kc_1 + c_2 + c_{10} = 0$$

$$(-k + 1)c_2 + c_1 = 0$$

$$(-k + 1)c_8 + c_9 = 0$$

$$-kc_9 + c_8 + c_{10} = 0$$

$$(-k + 1)c_{10} + c_1 + c_9 = 0$$

A_x: $c_1 = -c_4$; $c_2 = -c_3$; $c_8 = -c_7$; $c_9 = -c_6$; $c_{10} = -c_5$

$$-kc_1 + c_2 + c_{10} = 0$$
$$(-k - 1)c_2 + c_1 = 0$$
$$(-k - 1)c_8 + c_9 = 0$$
$$-kc_9 + c_8 + c_{10} = 0$$
$$(-k - 1)c_{10} + c_1 + c_9 = 0$$

2. Using the σ_y – plane of symmetry:

S_y: $c_1 = c_9$; $c_2 = c_8$; $c_3 = c_7$; $c_4 = c_6$; $c_5 = c_5$; $c_{10} = c_{10}$

$$-kc_1 + c_2 + c_{10} = 0$$
$$-kc_2 + c_1 + c_3 = 0$$
$$-kc_3 + c_2 + c_4 = 0$$
$$-kc_4 + c_3 + c_5 = 0$$
$$-kc_5 + 2c_4 + c_{10} = 0$$
$$-kc_{10} + 2c_1 + c_5 = 0$$

A_y: $c_1 = -c_9$; $c_2 = -c_8$; $c_3 = -c_7$; $c_4 = -c_6$; $c_5 = c_{10} = 0$

$$-kc_1 + c_2 = 0$$
$$-kc_2 + c_1 + c_3 = 0$$
$$-kc_3 + c_2 + c_4 = 0$$
$$-kc_4 + c_3 = 0$$

3. Using the σ_x – and σ_y – planes of symmetry:

$S_x S_y$: $c_1 = c_4 = c_6 = c_9$; $c_2 = c_3 = c_7 = c_8$; $c_{10} = c_5$

$$-kc_1 + c_2 + c_{10} = 0$$
$$(-k + 1)c_2 + c_1 = 0$$
$$(-k + 1)c_{10} + 2c_1 = 0$$

S_xA_y: $c_1 = c_4 = -c_6 = -c_9$; $c_2 = c_3 = -c_7 = -c_8$; $c_{10} = c_5 = 0$

$$-kc_1 + c_2 = 0$$
$$(-k + 1)c_2 + c_1 = 0$$

A_xS_y: $c_1 = -c_4 = -c_6 = c_9$; $c_2 = -c_3 = -c_7 = c_8$; $c_{10} = -c_5$

$$-kc_1 + c_2 + c_{10} = 0$$
$$(-k - 1)c_2 + c_1 = 0$$
$$(-k - 1)c_{10} + 2c_1 = 0$$

A_xA_y: $c_1 = -c_4 = c_6 = -c_9$; $c_2 = -c_3 = c_7 = -c_8$; $c_{10} = c_5 = 0$

$$-kc_1 + c_2 = 0$$
$$(-k - 1)c_2 + c_1 = 0$$

14.   (a) $k^2 - 1 = 0$ $(1; -1)$

       (b) $k^3 - 2k = 0$ $(\sqrt{2}; 0; -\sqrt{2})$

       (c) $k^4 - 3k^2 + 1 = 0$ $(1.618; 0.618; -0.618; -1.618)$

       (d) $-k^3 + 3k + 2 = 0$ $(1; 1; -2)$

       (e) $k^4 - 4k^2 = 0$ $(2; 0; 0; -2)$

15.   Result: $k^4 - k^3 - 2.75k^2 + 1.5k + 0.75 = 0$

       Procedure: $\alpha_N = \alpha + 0.5\beta$
                             $\beta_{CN} = \beta$
                             $\beta_{NN} = \beta$

                             1   2   3   4
                             N ———— N ———— C ———— C

$$(-k + 0.5)c_1 + c_2 = 0$$

$$(-k + 0.5)c_2 + c_1 + c_3 = 0$$

$$-kc_3 + c_2 + c_4 = 0$$

$$-kc_4 + c_3 = 0$$

$$D = \begin{vmatrix} -k + 0.5 & 1 & 0 & 0 \\ 1 & -k + 0.5 & 1 & 0 \\ 0 & 1 & -k & 1 \\ 0 & 0 & 1 & -k \end{vmatrix} =$$

$$(-k + 0.5) \begin{vmatrix} -k + 0.5 & 1 & 0 \\ 1 & -k & 1 \\ 0 & 1 & -k \end{vmatrix} \quad \begin{vmatrix} -k + 0.5 & 1 \\ 1 & -k \\ 0 & 1 \end{vmatrix} -D_1$$

where

$$-D_1 = - \begin{vmatrix} 1 & 0 & 0 \\ 1 & -k & 1 \\ 0 & 1 & -k \end{vmatrix} = -k^2 + 1$$

$$D = (-k + 0.5) \left\{ -k^3 + 0.5k^2 + k - 0.5 + k \right\} -D_1 =$$

$$= k^4 - 0.5k^3 - k^2 + 0.5k - k^2 - 0.5k^3 + 0.25k^2 +$$

$$+ 0.5k - 0.25 + 0.5k - D_1 =$$

$$= k^4 - k^3 - 2.75k^2 + 1.5k + 0.75 = 0$$

16.

S: $c_1 = c_5$; $c_2 = c_6$; $c_3 = c_7$; $c_4 = c_8$; $c_9 = c_{10}$

A: $c_1 = -c_5$; $c_2 = -c_6$; $c_3 = -c_7$; $c_4 = -c_8$; $c_9 = -c_{10}$

S:
$$-kc_1 + c_2 + c_9 = 0$$
$$(-k + 0.5)c_2 + c_1 + c_3 = 0$$
$$-kc_3 + c_2 + c_4 = 0$$
$$-kc_4 + c_3 + c_9 = 0$$
$$(-k + 1)c_9 + c_1 + c_4 = 0$$

A:
$$-kc_1 + c_2 + c_9 = 0$$
$$(-k + 0.5)c_2 + c_1 + c_3 = 0$$
$$-kc_3 + c_2 + c_4 = 0$$
$$-kc_4 + c_3 - c_9 = 0$$
$$(-k - 1)c_9 + c_1 - c_4 = 0$$

17.   (a)

$$\overset{1}{H_2C} = \overset{2}{N} - \overset{3}{CH} = \overset{4}{CH_2}$$

$$-kc_1 + c_2 = 0$$
$$(-k + 0.5)c_2 + c_1 + c_3 = 0$$
$$-kc_3 + c_2 + c_4 = 0$$
$$-kc_4 + c_3 = 0$$

$$\begin{vmatrix} -k & 1 & 0 & 0 \\ 1 & -k+0.5 & 1 & 0 \\ 0 & 1 & -k & 1 \\ 0 & 0 & 1 & -k \end{vmatrix} \qquad \begin{pmatrix} 0 & 1 & 0 & 0 \\ 1 & 0.5 & 1 & 0 \\ 0 & 1 & 0 & 1 \\ 0 & 0 & 1 & 0 \end{pmatrix}$$

(b)

$$(-k + 1)c_1 + 0.6c_2 + 0.6c_5 = 0$$
$$(-k + 0.5)c_2 + 0.6c_1 + c_3 = 0$$
$$-kc_3 + c_2 + c_4 = 0$$
$$-kc_4 + c_3 + c_5 = 0$$
$$-kc_5 + 0.6c_1 + c_4 = 0$$

$$\begin{vmatrix} -k+1 & 0.6 & 0 & 0 & 0.6 \\ 0.6 & -k+0.5 & 1 & 0 & 0 \\ 0 & 1 & -k & 1 & 0 \\ 0 & 0 & 1 & -k & 1 \\ 0.6 & 0 & 0 & 1 & -k \end{vmatrix}$$

$$\begin{pmatrix} 1 & 0.6 & 0 & 0 & 0.6 \\ 0.6 & 0.5 & 1 & 0 & 0 \\ 0 & 1 & 0 & 1 & 0 \\ 0 & 0 & 1 & 0 & 1 \\ 0.6 & 0 & 0 & 1 & 0 \end{pmatrix}$$

(c)

$$(-k+1.3)c_1 + \sqrt{2}\, c_2 = 0$$
$$(-k+0.2)c_2 + \sqrt{2}\, c_1 + c_3 + c_8 = 0$$
$$-kc_3 + c_2 + c_4 = 0$$
$$-kc_4 + c_3 + c_5 = 0$$
$$(-k+0.2)c_5 + c_4 + \sqrt{2}\, c_6 + c_7 = 0$$
$$(-k+1.3)c_6 + \sqrt{2}\, c_5 = 0$$
$$-kc_7 + c_5 + c_8 = 0$$
$$-kc_8 + c_2 + c_7 = 0$$

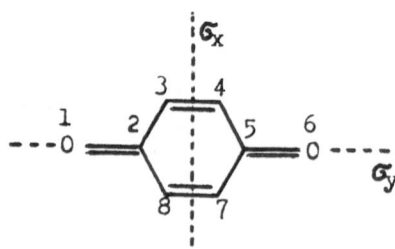

$$\begin{vmatrix} -k + 1.3 & \sqrt{2} & 0 & 0 & 0 & 0 & 0 & 0 \\ \sqrt{2} & -k + 0.2 & 1 & 0 & 0 & 0 & 0 & 1 \\ 0 & 1 & -k & 1 & 0 & 0 & 0 & 0 \\ 0 & 0 & 1 & -k & 1 & 0 & 0 & 0 \\ 0 & 0 & 0 & 1 & -k +0.2 & \sqrt{2} & 1 & 0 \\ 0 & 0 & 0 & 0 & \sqrt{2} & -k +1.3 & 0 & 0 \\ 0 & 0 & 0 & 0 & 1 & 0 & -k & 1 \\ 0 & 1 & 0 & 0 & 0 & 0 & 1 & -k \end{vmatrix}$$

$$\begin{pmatrix} 1.3 & \sqrt{2} & 0 & 0 & 0 & 0 & 0 & 0 \\ \sqrt{2} & 0.2 & 1 & 0 & 0 & 0 & 0 & 1 \\ 0 & 1 & 0 & 1 & 0 & 0 & 0 & 0 \\ 0 & 0 & 1 & 0 & 1 & 0 & 0 & 0 \\ 0 & 0 & 0 & 1 & 0.2 & \sqrt{2} & 1 & 0 \\ 0 & 0 & 0 & 0 & \sqrt{2} & 1.3 & 0 & 0 \\ 0 & 0 & 0 & 0 & 1 & 0 & 0 & 1 \\ 0 & 1 & 0 & 0 & 0 & 0 & 1 & 0 \end{pmatrix}$$

1. Using the σ_x-plane of symmetry:

S_x: $c_1 = c_6$; $c_2 = c_5$; $c_3 = c_4$; $c_7 = c_8$

$$(-k + 1.3)c_1 + \sqrt{2}\, c_2 = 0$$
$$(-k + 0.2)c_2 + \sqrt{2}\, c_1 + c_3 + c_8 = 0$$
$$(-k + 1)c_3 + c_2 = 0$$
$$(-k + 1)c_8 + c_2 = 0$$

A_x: $c_1 = -c_6$; $c_2 = -c_5$; $c_3 = -c_4$; $c_7 = -c_8$

$$(-k + 1.3)c_1 + \sqrt{2}\, c_2 = 0$$
$$(-k + 0.2)c_2 + \sqrt{2}\, c_1 + c_3 + c_8 = 0$$
$$(-k - 1)c_3 + c_2 = 0$$
$$(-k - 1)c_8 + c_2 = 0$$

2. Using the σ_y-plane of symmetry:

S_y: $c_1 = c_1$; $c_2 = c_2$; $c_5 = c_5$; $c_6 = c_6$; $c_3 = c_8$; $c_4 = c_7$

$$(-k + 1.3)c_1 + \sqrt{2}\, c_2 = 0$$
$$(-k + 0.2)c_2 + \sqrt{2}\, c_1 + 2c_3 = 0$$
$$-kc_3 + c_2 + c_4 = 0$$
$$-kc_4 + c_3 + c_5 = 0$$
$$(-k + 0.2)c_5 + 2c_4 + \sqrt{2}\, c_6 = 0$$
$$(-k + 1.3)c_6 + \sqrt{2}\, c_5 = 0$$

A_y: $c_1 = c_2 = c_5 = c_6 = 0$; $c_3 = -c_8$; $c_4 = -c_7$

$$-kc_3 + c_4 = 0$$
$$-kc_4 + c_3 = 0$$

3. Using the σ_x- and σ_y-planes of symmetry:

$S_x S_y$: $c_1 = c_6$; $c_2 = c_5$; $c_3 = c_4 = c_8 = c_7$

$$(-k + 1.3)c_1 + \sqrt{2}\, c_2 = 0$$
$$(-k + 0.2)c_2 + \sqrt{2}\, c_1 + 2c_3 = 0$$
$$(-k + 1)c_3 + c_2 = 0$$

$S_x A_y$: $c_1 = c_2 = c_5 = c_6 = 0$; $c_3 = c_4 = -c_8 = -c_7$

$$(-k + 1)c_3 = 0$$

$A_x S_y$: $c_1 = c_1$; $c_2 = c_2$; $c_3 = -c_4 = c_8 = -c_7$;
$c_5 = c_5$; $c_6 = c_6$

$$(-k + 1.3)c_1 + \sqrt{2}\, c_2 = 0$$
$$(-k + 0.2)c_2 + \sqrt{2}\, c_1 + 2c_3 = 0$$
$$(-k - 1)c_3 + c_2 = 0$$

$$A_x A_y: \quad c_1 = c_2 = c_5 = c_6 = 0; \quad c_3 = -c_4 = c_7 = -c_8$$

$$(-k - 1)c_3 = 0$$

18. The secular determinant must be symmetric with respect to the principal diagonal, this condition is not satisfied in this case.

19. The secular determinant concerns the $2p_z$ atomic orbitals and not the electrons which occupy these orbitals; it is, therefore, identical for the three forms given.

20.

S: $\quad (-k + \delta)c_1 + 2\rho c_2 = 0$

$\quad\quad -kc_2 + \rho c_1 + c_3 = 0$

$\quad\quad (1 - k)c_3 + c_2 = 0$

$$\begin{vmatrix} -k + \delta & 2\rho & 0 \\ \rho & -k & 1 \\ 0 & 1 & -k + 1 \end{vmatrix} =$$

$$= (-k + \delta)(k^2 - k) + k - \delta + 2\rho^2 k - 2\rho^2 =$$

$$= -k^3 + k^2 - k\delta + k^2 + k - \delta + 2\rho^2 k - 2\rho^2 =$$

$$= -k^3 + k^2(1 + \delta) + k(1 - \delta + 2\rho^2) - \delta - 2\rho^2 = 0$$

A: $\quad\quad\quad -kc_2 + c_3 = 0$

$\quad\quad -kc_3 + c_2 - c_3 = 0$

$$\begin{vmatrix} -k & 1 \\ 1 & -k - 1 \end{vmatrix} = k^2 + k - 1 = 0$$

The equation for the antisymmetric states (A) does not depend on δ and ρ.

Table of Parameters

Para-meter	δ	ρ	Equations for symmetric states
(a)	0.0	0.6	$-k^3 + k^2 + 1.72k - 0.72 = 0$
(b)	0.5	0.6	$-k^3 + 1.5k^2 + 1.22k - 1.22 = 0$
(c)	1.0	0.6	$-k^3 + 2k^2 + 0.72k - 1.72 = 0$
(d)	1.0	0.8	$-k^3 + 2k^2 + 1.28k - 2.28 = 0$
(e)	1.0	1.0	$-k^3 + 2k^2 + 2.00k - 3.00 = 0$

21. We tabulate the function $f(k) = k^4 - 4k^2 - 2k + 1$

k	f(k)	k	f(k)
2.2	0.665	0	1.000
2.0	-3.000	-0.1	1.160
1.9	-4.210	-0.2	1.242
1.7	-5.608	-0.3	1.248
1.5	-5.938	-0.4	1.180
1.4	-5.798	-0.6	0.889
1.3	-5.504	-0.8	0.449
1.0	-4.000	-0.9	0.216
0.9	-3.384	-1.0	0.000
0.8	-2.550	-1.2	-0.286
0.6	-1.510	-1.3	-0.316
0.4	-0.414	-1.4	-0.198
0.2	0.440	-1.5	0.062
0.1	0.760	-1.6	0.514
		-1.7	1.192
		-1.9	3.390
		-2.0	5.000

$$k = 2.17; \ 0.31; \ -1.00; \ -1.48$$

For graph, see Fig. 7.

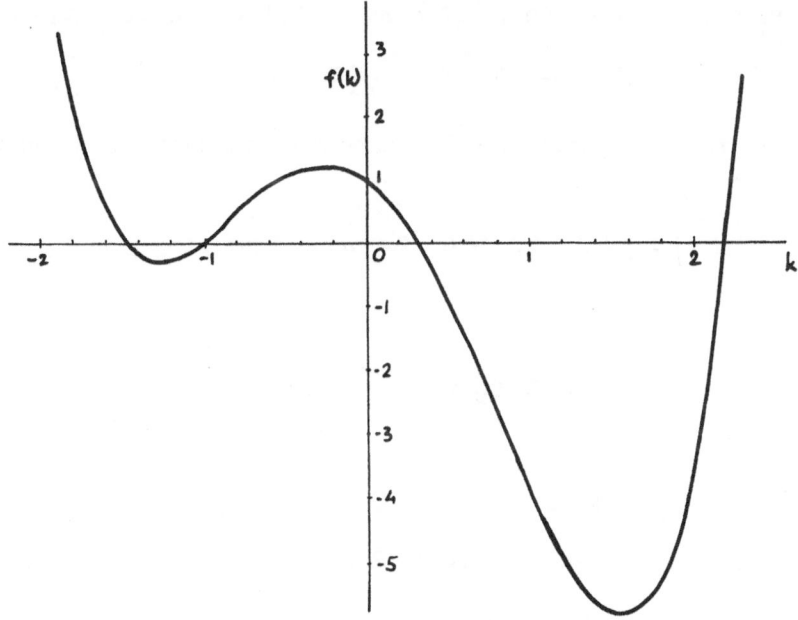

Fig. 7. Graphical solution of the secular polynomial for methylenecyclopropene (Exercise 21).

22. Use Equation (18) of the Appendix

$$E_1 = \frac{\int \varphi_1 H \varphi_1 d\tau}{\int \varphi_1^2 \, d\tau} \qquad (18)$$

φ_1 is normalized:

$$E_1 = \int \varphi_1 H \varphi_1 d\tau$$

$$E_1 = \int (0.300\, \chi_1 + 0.231\, \chi_2 + 0.231\, \chi_3 + 0.300\, \chi_4 +$$
$$+ 0.300\, \chi_5 + 0.231\, \chi_6 + 0.231\, \chi_7 + 0.300\, \chi_8 +$$
$$+ 0.461\, \chi_9 + 0.461\, \chi_{10})\ H\ (0.300\, \chi_1 + 0.231\, \chi_2 +$$
$$+ 0.231\, \chi_3 + 0.300\, \chi_4 + 0.300\, \chi_5 + 0.231\, \chi_6 +$$
$$+ 0.231\, \chi_7 + 0.300\, \chi_8 + 0.461\, \chi_9 + 0.461\, \chi_{10})\ d\tau$$

$$E_1 = 0.300 \times 0.300 \int \chi_1 H \chi_1 d\tau + 0.231 \times 0.300 \int \chi_2 H \chi_1 d\tau +$$
$$+ 0.231 \times 0.300 \int \chi_3 H \chi_1 d\tau + \cdots$$

The resonance integrals for nonneighboring atoms are considered to be equal to zero.

$$\int \chi_\mu H \chi_\nu \, d\tau = 0 \quad \text{for nonneighboring } \mu \text{ and } \nu ,$$

otherwise

$$\int \chi_\mu H \chi_\nu \, d\tau = \begin{cases} \alpha \; (\mu = \nu) \\ \beta \text{ neighboring } \mu \neq \nu \end{cases}$$

$$E_1 = \alpha + 2 \Big[4(0.300 \times 0.231) + 4(0.300 \times 0.461) +$$
$$+ (0.461 \times 0.461) \Big] \beta$$

$$E_1 = \alpha + 2.299 \, \beta$$

Expansion coefficients can be introduced directly into Equation (19), which one can easily get from Equation (18) (see Appendix)

$$E_1 = \alpha + 2 \sum_{\mu < \nu} \sum c_{1\mu} c_{1\nu} \beta_{\mu\nu}$$

$$(\beta_\mu \neq 0 \text{ for neighboring } \mu \text{ and } \nu)$$

23.

$$E_i = \frac{\int \varphi_i H \varphi_i d\tau}{\int \varphi_i^2 \, d\tau}$$

$$E_i = \sum_\mu c_{i\mu}^2 \alpha_\mu + 2 \sum_{\mu < \nu} \sum c_{i\mu} c_{i\nu} \beta_{\mu\nu}$$

$$dE_i \cong \sum c_{i\mu}^2 \, d\alpha_\mu$$

$$E_i' = E_i + \sum_\mu c_{i\mu}^2 \, \triangle \alpha_\mu$$

1-azabutadiene HN=CH—CH=CH$_2$

$$E_1 = 1.618 + (0.372)^2 \times 0.5 = 1.687$$
$$E_2 = 0.618 + (0.601)^2 \times 0.5 = 0.798$$
$$E_3 = -0.618 + (0.601)^2 \times 0.5 = -0.438$$
$$E_4 = -1.618 + (0.372)^2 \times 0.5 = -1.549$$

1-aminobutadiene H$_2$N—CH=CH—CH=CH$_2$

$$E_1 = 1.732 + (0.288)^2 \times 1 = 1.815$$
$$E_2 = 1.000 + (0.500)^2 \times 1 = 1.250$$
$$E_3 = 0.000 + (0.576)^2 \times 1 = 0.332$$
$$E_4 = -1.000 + (0.500)^2 \times 1 = -0.750$$
$$E_5 = -1.732 + (0.288)^2 \times 1 = -1.649$$

24. Antisymmetric states of m-quinodimethane (I) in fact represent two separate allyl systems, so that the orbital energies of this

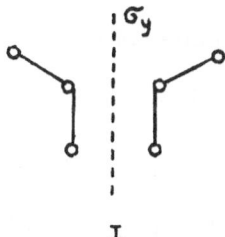

I

system are also the orbital energies of I. We shall now concentrate on the allyl unit; it is an odd alternant system, so that $k_1 = 0$ and $k_2 = -k_3$; in view of Equation (35) of the Appendix, we have

$$2 k_2^2 = 4$$
$$k_2 = \sqrt{2}; \quad k_3 = -\sqrt{2}$$

The energies α, $\alpha + \sqrt{2}\beta$, and $\alpha - \sqrt{2}\beta$ are also the orbital energies of I.

25. For antisymmetric states with respect to the σ_y-plane of symmetry the coefficients of atomic orbitals 9 and 10 are zero; clearly the orbital energies of benzene $(2, 1, 1, -1, -1, -2)$ are also the orbital energies of the three substances mentioned.

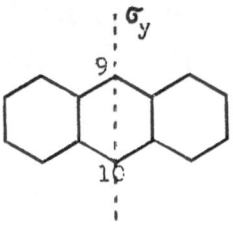

26. Use Equations (34) and (35) of the Appendix for the computation

$$E_5 = \alpha - 0.254$$

$$E_6 = \alpha + \beta$$

27.   (a)

$$k = 1.848; \ 0.765; \ 0.000; \ -0.765; \ -1.848$$

We are dealing with an odd alternant hydrocarbon ($k_4 = -0.765$; $k_3 = 0$). The sum of the squares of the eigenvalues, $\sum_i k_i^2$, is equal to double the number of $C-C$ bonds (compare Equation (35)]

$$(0.765)^2 + (-0.765)^2 + (k_1)^2 + (-k_1)^2 = 8$$

$$k_1 = + \sqrt{3.414} = 1.848$$

$$k_5 = -k_1 = -\sqrt{3.414} = -1.848$$

(b)

$$k = 2, \ \sqrt{2}, \ 0, \ 0, \ 0, \ -\sqrt{2}, \ -2$$

Situation for the A_x state (the nonzero coefficients are marked by circles):

We obtain four isolated atomic orbitals:

$$k_1 = k_2 = 0$$

Since we are dealing with an odd alternant system, the number of nonbonding MO's must be odd, so that

$$k_3 = 0$$

Situation for the A_y state:

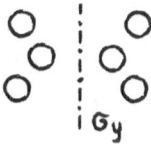

We get two allyl systems; the orbital energies of these systems are, therefore, also the orbital energies of the system considered: $k_4 = \sqrt{2}$ and, of course, also $k_5 = -\sqrt{2}$. (We only made use of the energies of orbitals which are symmetric with respect to the x-axis; we have already made use of the antisymmetric ones.)

The following holds for the remaining two orbital energies:

$$k_6 = -k_7$$

Use Equation (35):

$$\sum_i k_i^2 = 12$$

$$(\sqrt{2})^2 + (-\sqrt{2})^2 + (k_6)^2 + (k_7)^2 = 12$$

$$k_6 = 2$$

$$k_7 = -2$$

Note: The number of nonbonding orbitals (n) is defined by the difference between the number of starred and unstarred atoms:

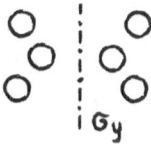

$$n = 5 - 2 = 3$$

This note indicates the alternative way of solving the problem.

28.

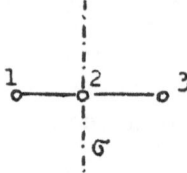

$$S: \quad -kc_1 + c_2 = 0$$
$$-kc_2 + 2c_1 = 0$$

$$\begin{vmatrix} -k & 1 \\ 2 & -k \end{vmatrix} = k^2 - 2; \quad k = \pm 1.414$$

$$A: \quad -kc_1 = 0$$
$$k = 0 \qquad (c_1 \neq 0)$$

Orbital energy:

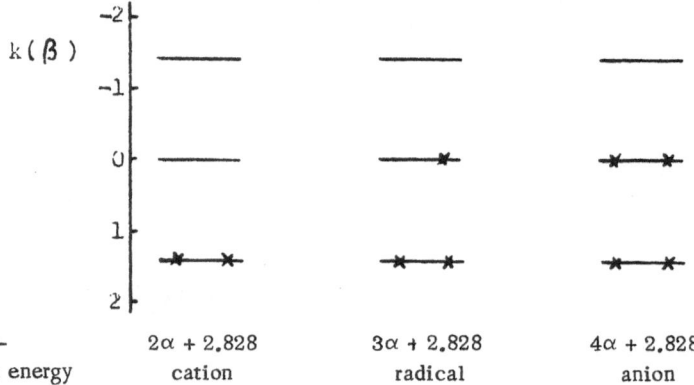

| Total π-electron energy | $2\alpha + 2.828$ cation | $3\alpha + 2.828$ radical | $4\alpha + 2.828$ anion |

Computation of coefficients:

$$S: \quad c_1' = c_1/c_2$$
$$c_2' = 1$$

$$-1.414 c_1' + 1 = 0$$
$$c_1' = \frac{1}{1.414}$$

Sym-metry	k	c_1'	c_2'	c_3'	$\sum c_i'^2$	N	c_1	c_2	c_3	c_1^2	c_2^2	c_3^2	$c_1 c_2$
S	1.414	$\frac{1}{1.414}$	1	$\frac{1}{1.414}$	2	$\frac{1}{\sqrt{2}}$	0.500	0.707	0.500	0.250	0.500	0.250	0.354
A	0	1	0	1	2	$\frac{1}{\sqrt{2}}$	0.707	0	0.707	0.500	0	0.500	0.000
S	-1.414	$-\frac{1}{1.414}$	$1 - \frac{1}{1.414}$	$-\frac{1}{1.414}$	2	$\frac{1}{\sqrt{2}}$	-0.500	0.707	-0.500	0.250	0.500	0.250	-0.354

Cation CH$_2$=CH–CH$_2$ (+)

$$
\begin{array}{ccc}
0.500 & 1.000 & 0.500 \\
\mathrm{C}\!-\!\!\!& \mathrm{C} &\!\!\!-\!\mathrm{C} \\
& 0.708 \quad 0.708 &
\end{array}
$$

$\sum q$
2.000

Radical CH$_2$=CH–CH$_2$

$$
\begin{array}{ccc}
1.000 & 1.000 & 1.000 \\
\mathrm{C}\!-\!\!\!& \mathrm{C} &\!\!\!-\!\mathrm{C} \\
& 0.708 \quad 0.708 &
\end{array}
$$

3.000

Anion CH$_2$=CH–CH$_2$ (−)

$$
\begin{array}{ccc}
1.500 & 1.000 & 1.500 \\
\mathrm{C}\!-\!\!\!& \mathrm{C} &\!\!\!-\!\mathrm{C} \\
& 0.708 \quad 0.708 &
\end{array}
$$

4.000

29. For systems with an axis of symmetry the lowest bonding molecular orbital is symmetric. We shall, therefore, use the system of secular equations for symmetric states. In order to obtain an inhomogeneous system of equations, we shall apply new coefficients:

$$c_1' = \frac{c_1}{c_1} = 1; \quad c_2' = \frac{c_2}{c_1}; \quad c_3' = \frac{c_3}{c_1}; \quad c_4' = \frac{c_4}{c_1}; \quad c_5' = \frac{c_5}{c_1}$$

$$
\begin{aligned}
-1.1c_2' + 2c_3' &= 1.1 \\
c_2' - 2.1c_3' + c_4' &= 0 \\
c_3' - 2.1c_4' + c_5' &= 0 \\
2c_4' - 2.1c_5' &= 0
\end{aligned}
$$

In order to determine the unknowns, it is necessary to compute the determinant of the system, Δ, as well as the determinants in which the right-hand side of the above equations is substituted for the i-th column:

$$
\Delta =
\begin{vmatrix}
-1.1 & 2 & 0 & 0 \\
1 & -2.1 & 1 & 0 \\
0 & 1 & -2.1 & 1 \\
0 & 0 & 2 & -2.1
\end{vmatrix}
= -1.1
\begin{vmatrix}
-2.1 & 1 & 0 \\
1 & -2.1 & 1 \\
0 & 2 & -2.1
\end{vmatrix}
-
\begin{vmatrix}
2 & 0 & 0 \\
1 & -2.1 & 1 \\
0 & 2 & -2.1
\end{vmatrix}
=
$$

$$= -1.1(-9.26 + 4.2 + 2.1) - (4.41 \times 2 - 4) = -1.56$$

$$
\Delta_2 =
\begin{vmatrix}
1.1 & 2 & 0 & 0 \\
0 & -2.1 & 1 & 0 \\
0 & 1 & -2.1 & 1 \\
0 & 0 & 2 & -2.1
\end{vmatrix}
= 1.1
\begin{vmatrix}
-2.1 & 1 & 0 \\
1 & -2.1 & 1 \\
0 & 2 & -2.1
\end{vmatrix}
=
$$

$$= 1.1(-9.26 + 4.2 + 2.1) = -3.256$$

$$\Delta_3 = \begin{vmatrix} -1.1 & 1.1 & 0 & 0 \\ 1 & 0 & 1 & 0 \\ 0 & 0 & -2.1 & 1 \\ 0 & 0 & 2 & -2.1 \end{vmatrix} = -1.1 \begin{vmatrix} 1 & 1 & 0 \\ 0 & -2.1 & 1 \\ 0 & 2 & -2.1 \end{vmatrix} =$$

$$= -1.1(4.41 - 2) = -2.651$$

$$\Delta_4 = \begin{vmatrix} -1.1 & 2 & 1.1 & 0 \\ 1 & -2.1 & 0 & 0 \\ 0 & 1 & 0 & 1 \\ 0 & 0 & 0 & -2.1 \end{vmatrix} = 1.1 \begin{vmatrix} 1 & -2.1 & 0 \\ 0 & 1 & 1 \\ 0 & 0 & -2.1 \end{vmatrix} =$$

$$= 1.1(-2.1) = -2.31$$

$$\Delta_5 = \begin{vmatrix} -1.1 & 2 & 0 & 1.1 \\ 1 & -2.1 & 1 & 0 \\ 0 & 1 & -2.1 & 0 \\ 0 & 0 & 2 & 0 \end{vmatrix} = -1.1 \begin{vmatrix} 1 & -2.1 & 1 \\ 0 & 1 & -2.1 \\ 0 & 0 & 2 \end{vmatrix} =$$

$$= -1.1 \times 2 = -2.2$$

We can now compute the unknowns:

$$c_2' = \frac{\Delta_2}{\Delta} = \frac{-3.256}{-1.56} = 2.09$$

$$c_3' = \frac{\Delta_3}{\Delta} = \frac{-2.651}{-1.56} = 1.70$$

$$c_4' = \frac{\Delta_4}{\Delta} = \frac{-2.31}{-1.56} = 1.48$$

$$c_5' = \frac{\Delta_5}{\Delta} = \frac{-2.2}{-1.56} = 1.41$$

The coefficients without the dash must satisfy the normalization condition:

$$c_1^2 + c_2^2 + 2c_3^2 + 2c_4^2 + c_5^2 = 1$$

$$1 + c_2'^2 + 2c_3'^2 + 2c_4'^2 + c_5'^2 = \frac{1}{c_1^2}$$

$$c_1 = \frac{1}{\sqrt{1 + c_2'^2 + 2c_3'^2 + 2c_4'^2 + c_5'^2}}$$

$$c_1 = \frac{1}{\sqrt{1 + 4.37 + 2 \times 2.89 + 2 \times 2.19 + 1.99}}$$

$$c_1 = 0.238$$

$$c_2 = c_2' \times c_1 = 2.09 \times 0.238 = 0.500$$

$$c_3 = c_3' \times c_1 = 1.70 \times 0.238 = 0.406$$

$$c_4 = c_4' \times c_1 = 1.48 \times 0.238 = 0.353$$

$$c_5 = c_5' \times c_1 = 1.41 \times 0.238 = 0.336$$

30. Butadiene, p-quinodimethane, and naphthalene are alternant hydrocarbons. Therefore, their orbital energies are symmetrically arranged about the value α, i.e., the energy values form pairs $\alpha \pm k_i \beta$. The coefficients of the i-th molecular orbital with an energy $\alpha - k_i \beta$ can be found from the coefficients of the molecular orbital with an energy $\alpha + k_i \beta$ by changing the sign of the coefficients of one class of atoms (either starred or unstarred).

	E (β)	Symmetry	c_1	c_2	c_3	c_4
φ_3	-0.61803	Sx	0.60150	-0.37175	-0.37175	0.60150
φ_4	-1.61803	Ax	0.37175	-0.60150	0.60150	-0.37175

p-Quinodimethane

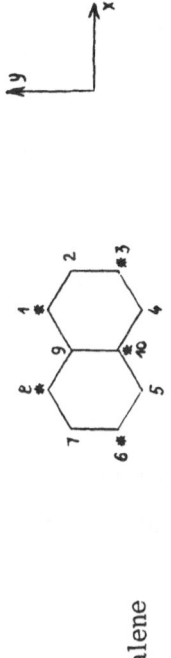

	E(β)	Sym-metry	c_1	c_2	c_3	c_4	c_5	c_6	c_7	c_8
φ_5	-0.31111	SxAy	0.17934	0.26033	-0.26033	-0.17934	-0.26033	0.26033	-0.57645	0.57648
φ_6	-1.00000	AxAy	0.00000	-0.50000	0.50000	0.00000	-0.50000	0.50000	0.00000	0.00000
φ_7	-1.48119	SxSy	0.52990	-0.21357	-0.21357	0.52990	-0.21357	-0.21357	-0.35775	-0.35775
φ_8	-2.17009	SxAy	0.43249	-0.36962	0.36962	-0.43249	0.36962	-0.36962	-0.19929	0.19929

Naphthalene

	E(β)	Sym-metry	c_1	c_2	c_3	c_4	c_5	c_6	c_7	c_8	c_9	c_{10}
φ_6	-0.61803	SxAy	0.42533	-0.26286	-0.26286	0.42533	-0.42533	0.26286	0.26286	-0.42533	0.00000	0.00000
φ_7	-1.00000	AxSy	0.00000	0.40825	-0.40825	0.00000	0.00000	-0.40825	0.40825	0.00000	-0.40825	0.40825
φ_8	-1.30278	SxSy	0.39958	-0.17352	-0.17352	0.39958	0.39958	-0.17352	-0.27352	0.39958	-0.34705	-0.34705
φ_9	-1.61803	AxAy	0.26287	-0.42533	0.42533	-0.26287	0.26287	-0.42533	0.42533	-0.26287	0.00000	0.00000
φ_{10}	-2.30278	AySy	0.30055	-0.23070	0.23070	-0.30055	-0.30055	0.23070	-0.23070	0.30055	-0.46140	0.46140

(a) For butadiene, we have

$$\varphi_3 = 0.60150\,\chi_1 - 0.37175\,\chi_2 - 0.37175\,\chi_3 + 0.60150\,\chi_4$$

$$\varphi_4 = 0.37175\,\chi_1 - 0.60150\,\chi_2 + 0.60150\,\chi_3 - 0.37175\,\chi_4$$

$$\int \varphi_3\,\varphi_4\,d\tau = \int (0.60150\,\chi_1 - 0.37175\,\chi_2 - 0.37175\,\chi_3 +$$
$$+ 0.60150\,\chi_4)(0.37175\,\chi_1 - 0.60150\,\chi_2 +$$
$$+ 0.60150\,\chi_3 - 0.37175\,\chi_4)\,d\tau$$

Since

$$\int \chi_\mu \chi_\nu d\tau = 0 \quad (\text{for } \mu \neq \nu)$$

we have

$$\int \varphi_3\,\varphi_4\,d\tau = (0.60150 \times 0.37175)\int \chi_1^2\,d\tau + (0.37175 \times$$
$$\times 0.60150) \times \int \chi_2^2\,d\tau - (0.37175 \times 0.60150) \times$$
$$\times \int \chi_3^2\,d\tau - (0.60150 \times 0.37175)\int \chi_4^2\,d\tau$$

The χ_μ are normalized; therefore

$$\int \chi_\mu^2\,d\tau = 1$$

$$\int \varphi_3\,\varphi_4\,d\tau = (0.60150 \times 0.37175) + (0.37175 \times 0.61050) -$$
$$- (0.37175 \times 0.61050) - (0.60150 \times 0.37175) = 0$$

For the other two systems, one may proceed analogously.

(b) For position 1 with naphthalene we have

$$c_{61}^2 + c_{71}^2 + c_{81}^2 + c_{91}^2 + c_{101}^2 = (0.42533)^2 + 0 + (0.39958)^2 +$$
$$+ (0.26287)^2 + (0.30055)^2 = 0.500$$

31. (a) $2\alpha + 2\beta$

  (b) 2α

  (c) 4α

32. (a) The computation may be carried out because

$$\sum_i k_i = 0, \qquad \sum_{\substack{\text{occupied} \\ \text{orbitals}}} k_i = -\sum_{\substack{\text{unoccupied} \\ \text{orbitals}}} k_i$$

(b) It is possible, provided the value of α_N

$$(\alpha_N = \alpha + \delta_N \beta)$$

is given for which the energies of the antibonding MO's were computed, because

$$\sum_i k_i = \delta_N$$

(c) Only for alternant hydrocarbons.

33.   I: $W = 10\alpha + 13.684\beta$ II: $W = 10\alpha + 13.364\beta$

       DE $= 3.684\beta$ DE $= 3.364\beta$

       $DE_n = 0.3684\beta$ $DE_n = 0.3364\beta$

       $DE_m = \dfrac{DE}{11} = 0.3349\beta$ $DE_m = \dfrac{DE}{11} = 0.3058\beta$

   $E(N \rightarrow V_1) = 1.236\beta$ $E(N \rightarrow V_1) = 0.477 + 0.400 = 0.877\beta$

34.   I: $W = 6\alpha + 8.074\beta$ II: $W = 6\alpha + 9.130\beta$

           DE $= 1.074\beta$ DE $= 1.130\beta$

           $DE_m = 0.215\beta$ $DE_m = 0.226\beta$

       $E(N \rightarrow V_1) = 1.626\beta$ $E(N \rightarrow V_1) = 1.566\beta$

       III: $W = 6\alpha + 7.418\beta$

           DE $= 1.418\beta$

           $DE_m = 0.283\beta$

       $E(N \rightarrow V_1) = 1.709\beta$

35. (a) Anthracene is an alternant hydrocarbon and, therefore, it has symmetrically arranged orbital energies in pairs, $E_i = \alpha \pm k_i\beta$. The orbital energies (expressed in terms of $-k$) of the unoccupied orbitals are as follows:

   $$-0.414; \; -1.000; \; -1.000; \; -1.414; \; -1.414; \; -2.000; \; -2.414$$

   (b) $W = 14\alpha + 19.312\beta$
   (c) $E(N \longrightarrow V_1) = 0.414 + 0.414 = 0.828\beta$
   (d) $DE = 5.312\beta$

   $\qquad DE = -20 \times 5.312 = -106.2$ kcal/mol

36. Use Equation (18) of the Appendix:

$$E = \int (0.6015\, \chi_1 + 0.3718\, \chi_2 - 0.3718\, \chi_3 - 0.6015\, \chi_4) \times$$

$$\times \; H \; (0.6015\, \chi_1 + 0.3718\, \chi_2 - 0.3718\, \chi_3 - 0.6015\, \chi_4)\, d\tau$$

$$E = 0.6015 \times 0.6015 \int \chi_1\, H\, \chi_1\, d\tau \; + \; 0.6015 \times 0.3718 \times$$

$$\times \int \chi_1\, H\, \chi_2\, d\tau - 0.6015 \times 0.3718 \int \chi_1\, H\, \chi_3\, d\tau + \ldots$$

$$E = \alpha + 2 \left[(2 \times 0.2236) - 0.1382 \right]\beta$$

The same expression is obtained if Equation (19) of the Appendix is used:

$$E_i = \alpha + 2 \sum_{\mu < \nu} \sum c_{i\mu}\, c_{i\nu}\, \beta_{\mu\nu} \qquad (\beta_{\mu\nu} \neq 0 \text{ if } \mu \text{ and } \nu \text{ are neighbors})$$

$$E = \alpha + 0.618\beta$$

the orbital energy of an antisymmetric bonding MO.

Total π-electron energy:

$$W = 6\alpha + 2 \left[2.1149 + 1.0000 + 0.6180 \right]\beta$$

$$W = 6\alpha + 7.4658\beta$$

37. Use Equation (28) of the Appendix:

$$\sum_\mu q_\mu = n \qquad\qquad 4q_1 + 2 \times 1.173 + 2.1198 = 8$$

$$q_1 = 0.815$$

38. Use Equations (28) and (20) of the Appendix:

$$0.622 + 1.047 + (2 \times 1.073) + 2q = 6$$

$$q = 1.092$$

$$W = \sum_{\mu} q_{\mu}\alpha + 2\sum_{\mu < \nu}\sum P_{\mu\nu}\,\beta_{\mu\nu}$$

$$6\alpha + 7.466\,\beta = 6\alpha + 2\left[0.759 + 2(0.778) + 0.520 + 2p\right]$$

$$p = 0.449$$

39. Use Equation (20) of the Appendix:

$$W = \sum_{\mu} q_{\mu}\alpha + 2\sum_{\mu < \nu}\sum P_{\mu\nu}\,\beta$$

$$10\alpha + 13.683\,\beta = 10\alpha + 2\left[4(0.555 + 0.725) + 2(0.608) + P_{910}\right]\beta$$

$$P_{910} = 0.515$$

$$DE = 3.683\,\beta$$

40. For the matrix of expansion coefficients of molecular orbitals

$$\begin{pmatrix} c_{11} & c_{12} & c_{13} & \cdots & c_{1\mu} \\ c_{21} & c_{22} & c_{23} & \cdots & c_{2\mu} \\ \vdots & & & & \\ c_{i1} & c_{i2} & c_{i3} & \cdots & c_{i\mu} \\ \vdots & & & & \\ c_{n1} & c_{n2} & c_{n3} & \cdots & c_{n\mu} \end{pmatrix}$$

where i refers to molecular orbitals and μ to atomic orbitals, we have

$$\sum_{i=1}^{n} c_{i\mu}^2 = 1 \tag{a}$$

$$\sum_{i=1}^{n} c_{i\mu} c_{i\nu} = 0 \tag{b}$$

In both cases the summation is carried out over all molecular orbitals (occupied and unoccupied).

From (a) it follows that

$$q_\mu = 2(1 - \sum_{unocc.} c_{i\mu}^2)$$

From (b) it follows that

$$P_{\mu\nu} = -2 \sum_{unocc.} c_{i\mu} c_{i\nu}$$

$$q_1 = 2 \{1 - [(-0.54591)^2 + 0 + (-0.32307)^2] \}$$

$$q_1 = 2 \{1 - 0.40238\}$$

$$q_1 = 1.1952$$

$$q_2 = 2 \{1 - [(0.36602)^2 + (0.5)^2 + (0.39313)^2]\} = 0.9230$$

$$q_3 = 2 \{1 - [(0.23810)^2 + (-0.5)^2 + (-0.43711)^2]\} = 1.0045$$

$$q_4 = 2 \{1 - [(-0.56626)^2 + 0 + (0.45210)^2]\} = 0.9499$$

$$P_{12} = -2 [(-0.54591)(0.36602) + (0.00000)(0.50000) +$$

$$+ (-0.32307)(0.39313)] = -2 [(-0.1998) + (-0.1270)] =$$

$$= 0.6537$$

$$P_{23} = -2 [(0.36602)(0.23810) + (-0.50000)(0.50000) +$$

$$+ (0.39313)(-0.43711)] = 0.6694$$

$$P_{34} = -2 \left[(-0.56626)(0.23810) + (-0.50000)(0.00000) + \right.$$
$$\left. + (-0.40711)(0.45210) \right] = 0.6649$$

The superdelocalizabilities for electrophilic substitution cannot be computed from the expansion coefficients of the antibonding molecular orbitals.

41. State due to 1 → 1' excitation

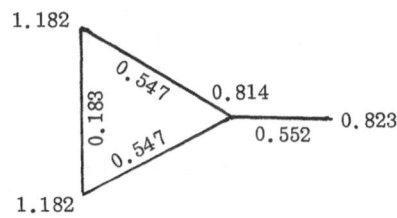

Computation of electron densities [see Equation (6) of the Appendix]:

$$q_1 = 2(0.282)^2 + (0.815)^2 = 0.823$$

$$q_2 = 2(0.612)^2 + (0.254)^2 = 0.814$$

$$q_3 = q_4 = 2(0.523)^2 + (0.368)^2 + (0.707)^2 = 1.182$$

Computation of bond orders [see Equation (8) of the Appendix]:

$$P_{12} = 2(0.282 \times 0.612) + (0.815 \times 0.254) = 0.552$$

$$P_{23} = P_{24} = 2(0.612 \times 0.523) - (0.254 \times 0.368) = 0.547$$

$$P_{34} = 2(0.523)^2 + (0.368)^2 - (0.707)^2 = 0.183$$

Computation of the total π-electron energy [see Equation (1) of the Appendix]:

$$W = 4\alpha + 2(2.170) + 0.311 - 1.000 = 4\alpha + 3.651\,\beta$$

Check of computed electron densities and bond orders [see Equation (20) of the Appendix]:

$$4\alpha + 3.651\beta \doteq \left[0.823 + 0.814 + (2 \times 1.182) \right] \alpha +$$
$$+ 2 \left[0.552 + (2 \times 0.547) + 0.183 \right] \beta$$

State due to 1 → 2' excitation

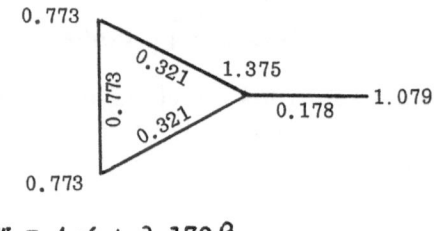

$$W = 4\alpha + 3.170\beta$$

Check of electron densities and bond orders:

$$4\alpha + 3.170\beta \doteq \left[1.079 + 1.375 + (2 \times 0.773)\right]\alpha +$$

$$+ 2\left[0.178 + (2 \times 0.321) + 0.773\right]\beta$$

42. Check of coefficients (for MO with an energy $E = \alpha + 2.303\beta$):

$$\sum_{i=1}^{10} c_i^2 = 1$$

$$4(0.30055)^2 + 4(0.23070)^2 + 2(0.46140)^2 = 1$$

Computation of electron density:

$$q_1 = \sum_{i=1}^{5} 2c_{i1}^2$$

$$q_1 = 2\left\{(0.30055)^2 + (0.26287)^2 + (0.39958)^2 + 0 + \right.$$

$$\left. + (0.42533)^2\right\}$$

$$q_1 = 1.000$$

Note: For alternant hydrocarbons in the ground state all electron densities are unity. The same is true also within the framework of the SCF method.

Computation of the bond order 1-2:

$$P_{12} = \sum_{i=1}^{5} 2c_{i1} \, c_{i2}$$

$$P_{12} = 2 \left\{ (0.30055 \times 0.23070) + (0.26287 \times 0.42533) + \right.$$

$$+ (0.39958 \times 0.17352) - (0 \times 0.40825) +$$

$$\left. + (0.42533 \times 0.26286) \right\}$$

$$P_{12} = 0.7246; \quad P_{19} = 0.5547$$

Computation of the free valence:

$$F_1 = \sqrt{3} - \sum_{k=2;9} P_{1k}$$

$$F_1 = \sqrt{3} - (0.7246 + 0.5547)$$

$$F_1 = 0.4528$$

Computation of the approximate superdelocalizability:

$$S_1' = 2 \, \frac{c_{51}^2}{k_5} = 2 \, \frac{(0.42533)^2}{0.61803}$$

$$S_1' = 0.585$$

Computation of the accurate superdelocalizability:

$$S_1 = 2 \sum_{i=1}^{5} \frac{c_{i1}^2}{k_i}$$

$$S_1 = 2 \left\{ \frac{(0.30055)^2}{2.30278} + \frac{(0.26287)^2}{1.61803} + \frac{(0.39958)^2}{1.30278} + \frac{0.00000}{1.00000} + \right.$$

$$\left. + \frac{(0.42533)^2}{0.61803} \right\}$$

$$S_1 = 0.994$$

43. Use the same equations as in Exercise 42 (see Appendix):

(a) $q_1 = 2\left[(0.42937)^2 + 0 + (0.60150)^2\right]$

   $q_1 = 1.092$

μ	1	2	5	6
q_μ	1.092	1.073	1.047	0.622

(b) $P_{12} = 2(0.42937 \times 0.38513 - 0 \times 0.50000 +$

   $+ 0.60150 \times 0.37175)$

   $P_{12} = 0.778$

$\mu\nu$	1-2	2-3	1-5	5-6
$P_{\mu\nu}$	0.778	0.520	0.449	0.759

(c) $F_1 = \sqrt{3} - (0.449 + 0.778)$

   $F_1 = 0.505$

μ	1	2	5	6
F_μ	0.505	0.434	0.075	0.973

(d) $S'_{1,e} = 2\,\dfrac{c_{31}^2}{k_3}$; $S'_{1,n} = -2\,\dfrac{c_{41}^2}{k_4}$; $S'_{1,r} = \dfrac{1}{2}\,(S'_{1,e} + S'_{1,n})$

[See Equations (11), (12), and (13) of the Appendix.]

$$S'_{1,e} = 2\,\frac{(0.60150)^2}{0.61805} = 1.17078$$

μ	$S'_{\mu,e}$	$S'_{\mu,n}$	$S'_{\mu,r}$
1	1.17078	0.96716	1.06897
2	0.44720	0.61496	0.53108
5	0.00000	0.28546	0.14273
6	0.00000	4.42126	2.21063

(e) Use Equations (11), (12), and (13):

$$S_{1,e} = 2 \sum_{i=1}^{3} \frac{c_{i1}^2}{k_i}$$

$$S_{1,e} = 2 \left[\frac{(0.42937)^2}{2.11492} + \frac{(0.00000)^2}{1.00000} + \frac{(0.60150)^2}{0.61805} \right]$$

$$S_{1,e} = 1.34512$$

μ	$S_{\mu,e}$	$S_{\mu,n}$	$S_{\mu,r}$
1	1.34512	1.34514	1.34513
2	1.08746	1.08748	1.08747
5	0.75862	0.75862	0.75862
6	0.55782	4.55786	2.55784

Check:

(a) π-electron densities:

$$\sum_{\mu=1}^{6} q_\mu = 6$$

$$(2 \times 1.092) + (2 \times 1.073) + 1.047 + 0.622 = 6$$

(b) Bond orders:

$$W = \sum_\mu q_\mu \alpha + 2 \sum_{\mu < \nu} \sum P_{\mu\nu} \beta_{\mu\nu}$$

We know that the total π-electron energy (obtained by adding up the orbital energies) amounts to

$$W = 6\alpha + 7.4659\beta$$

$$6\alpha + 7.4659\beta = 6\alpha + 2\sum_{\mu < \nu}\sum P_{\mu\nu}\beta_{\mu\nu}$$

$$7.4659\beta = \left[(2 \times 0.778) + 0.520 + (2 \times 0.449) + \right.$$
$$\left. + 0.759\right]\beta$$

(c) Free valences

Use Equation (29) of the Appendix:

$$\sum_{\mu=1}^{6} F_{\mu} = 6\sqrt{3} - \frac{W - 6\alpha}{\beta}$$

$$(2 \times 0.505) + 2(0.434) + 0.075 + 0.973 = 6\sqrt{3} - 7.4659$$

(d) Approximate superdelocalizabilities

Use Equations (32) and (33) of the Appendix:

$$\sum_{\mu} S'_{\mu, e} = 2\sum_{\mu} \frac{c^2_{3\mu}}{k_3}$$

$$\sum_{\mu} S'_{\mu, e} = \frac{2}{k_3}$$

$$(2 \times 1.17078) + (2 \times 0.44720) = \frac{2}{0.61805}$$

$$3.2359 = 3.2359$$

$$\sum_{\mu} S'_{\mu, n} = -2\sum_{\mu} \frac{c^2_{4\mu}}{k_4}$$

$$\sum_{\mu} S'_{\mu, n} = -\frac{2}{k_4}$$

$$(2 \times 0.96716) + (2 \times 0.61496) + 0.28546 + 4.42126 = \frac{2}{0.2541}$$

$$7.8709 = 7.8709$$

$$\sum_{\mu} S'_{\mu,r} = \sum_{\mu} \frac{1}{2} (S'_{\mu,e} + S'_{\mu,n})$$

$$\sum_{\mu} S'_{\mu,n} = \frac{1}{k_3} - \frac{1}{k_4}$$

(e) Accurate superdelocalizabilities

Use Equations (30) and (31) of the Appendix:

$$\sum_{\mu} S_{\mu,e} = 2 \sum_{\mu=1}^{6} \sum_{i=1}^{3} \frac{c_{i\mu}^2}{k_i} = 2 \sum_{\mu=1}^{3} \frac{1}{k_i}$$

$(2 \times 1.34512) + 2(1.08746) + 0.75862 + 0.55782 =$

$$= \frac{2}{2.11492} + \frac{2}{1.00000} + \frac{2}{0.61805}$$

$6.1816 = 0.94565 + 2 + 3.23598$

$6.1816 = 6.1816$

$$\sum_{\mu} S_{\mu,n} = -2 \sum_{\mu=1}^{6} \sum_{i=4}^{6} \frac{c_{i\mu}^2}{k_i} = -2 \sum_{i=4}^{6} \frac{1}{k_i}$$

$(2 \times 1.34514) + (2 \times 1.08748) + 0.75862 + 4.55786 =$

$$= \frac{2}{0.25410} + \frac{2}{1.61803} + \frac{2}{1.86081}$$

$10.1817 = 7.87091 + 1.23607 + 1.07480$

$10.1817 = 10.1817$

Pyridine

$$\begin{array}{c} 4 \\ 3 \quad \quad 5 \\ 2 \quad N \quad 6 \\ 1 \end{array}$$

μ	q	F	S'_e	S'_r	S'_n	S_e
1	1.1952	–	0.0000	0.3543	0.7087	0.8167
2	0.9230	0.4090	0.5000	0.1593	0.3186	0.7285
3	1.0045	0.3978	0.5000	0.0674	0.1348	0.8324
4	0.9499	0.4023	0.0000	0.3813	0.7626	0.7240

μ	S_r	S_n
1	0.8167	0.8167
2	0.8505	0.9785
3	0.8324	0.8324
4	0.8490	0.9740

$\mu\nu$	$P_{\mu\nu}$
1-2	0.6537
2-3	0.6694
3-4	0.6649

Check:

(a) π-electron densities:

$$\sum_{\mu=1}^{6} q_\mu = 6$$

$$1.1952 + (2 \times 0.9230) + (2 \times 1.0045) + 0.9499 = 6$$

(b) Bond orders [see Equation (20) of the Appendix]:

We know that the total π-electron energy, computed from the orbital energies, is

$$W = 6\alpha + 8.54928\,\beta$$

$$W = 6\alpha + q_N\,0.5\,\beta + 2\sum_{\mu<\nu}\sum P_{\mu\nu}\,\beta_{\mu\nu}$$

$$W = 6\alpha + 8.549\,\beta$$

(c), (d), (e) Check of free valences and superdelocalizabilities as with fulvene.

44.

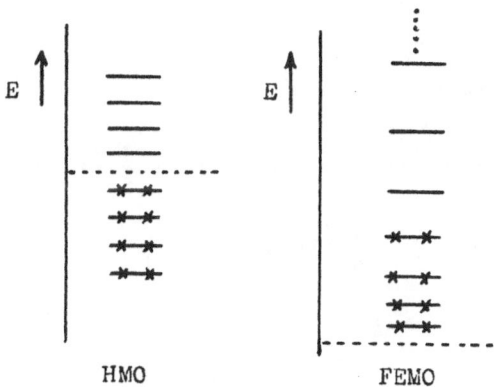

HMO   FEMO

The π-electronic system of octatetraene consists of eight $2p_z$ atomic orbitals. This means that with the HMO method we obtain eight π-molecular orbitals. The FEMO method leads to an infinite number of molecular orbitals. In both cases the four lowest molecular orbitals are occupied.

45*. There are three possible structures:

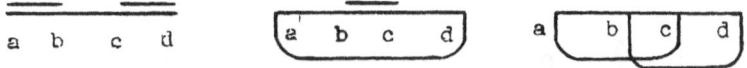

The functions appropriate to these structures, however, are not linearly independent, i.e., one may express one function by a linear combination of the other two. The total wave function of butadiene will, therefore, be formed from two of the three structures mentioned above. Let us select

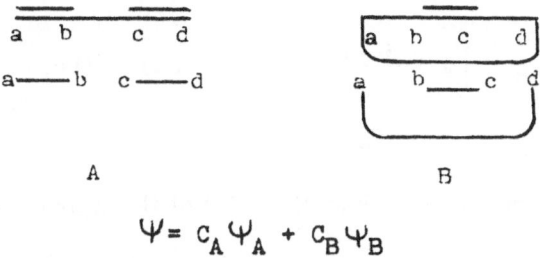

A   B

$$\Psi = c_A \Psi_A + c_B \Psi_B$$

---

*A detailed description of the VB procedure is given in the Appendix.

H_{AA}, S_{AA}

Cycles $(\frac{a}{b})$, $(\frac{c}{d})$, $x = 2$, $y = 2$

$$H_{AA} = \frac{2^2}{2^2}\left[Q + \frac{3}{2}\left\{(ab) + (cd) - 0\right\} - \frac{1}{2}\left\{(ab) + (bc) + (cd)\right\}\right]$$

$$H_{AA} = Q + \frac{3}{2}(2\alpha) - \frac{1}{2}(3\alpha) \tag{45.1}$$

$$H_{AA} = Q + \frac{3}{2}\alpha \qquad S_{AA} = 1 \tag{45.2}$$

H_{AB}, S_{AB}

Cycle $(\frac{ac}{bd})$ $x = 1$, $y = 2$

$$H_{AB} = \frac{2}{2^2}\left[Q + \frac{3}{2}\left\{(ab) + (bc) + (cd)\right\} - \frac{1}{2}\left\{(ab) + (bc) + (cd)\right\}\right]$$

$$H_{AB} = \frac{1}{2}\left[Q + \frac{3}{2}(3\alpha) - \frac{1}{2}(3\alpha)\right]$$

$$H_{AB} = \frac{1}{2}Q + \frac{3}{2}\alpha \qquad S_{AB} = \frac{1}{2} \tag{45.3}$$

H_{BB}, S_{BB}

Cycles $(\frac{a}{d})$, $(\frac{b}{c})$ $x = 2$, $y = 2$

$$H_{BB} = \left[Q + \frac{3}{2}\left\{(bc)\right\} - \frac{1}{2}\left\{(ab) + (bc) + (cd)\right\}\right]$$

$$H_{BB} = Q \qquad S_{BB} = 1 \tag{45.4}$$

We shall substitute into Equation (44) of the Appendix:

$$C_A\left[Q + \frac{3}{2}\alpha - E\right] + C_B\left[\frac{1}{2}Q + \frac{3}{2}\alpha - \frac{1}{2}E\right] = 0$$

$$C_A\left[\frac{1}{2}Q + \frac{3}{2}\alpha - \frac{1}{2}E\right] + C_B\left[Q - E\right] = 0 \tag{45.5}$$

Secular determinant:

$$\begin{vmatrix} Q + \frac{3}{2}\alpha - E & \frac{1}{2}Q + \frac{3}{2}\alpha - \frac{1}{2}E \\ \frac{1}{2}Q + \frac{3}{2}\alpha - \frac{1}{2}E & Q - E \end{vmatrix} = 0 \qquad (45.6)$$

Solution:

The value

$$E = Q \pm \sqrt{3}\,\alpha \qquad (45.7)$$

is negative; the ground state thus has the energy

$$E = Q + \sqrt{3}\,\alpha \qquad (45.8)$$

The resonance energy is the difference between the π-electron energy of the ground state and the energy of the structure A:

$$E_A = \frac{\int \Psi_A H \Psi_A \, d\tau}{\int \Psi_A \Psi_A \, d\tau} = \frac{H_{AA}}{S_{AA}} = Q + \frac{3}{2}\alpha \qquad (45.9)$$

$$E - E_A = Q + \sqrt{3}\,\alpha - Q - \frac{3}{2}\alpha = (\sqrt{3} - \frac{3}{2})\alpha = 0.232\,\alpha$$

Computation of coefficients:

We substitute (45.8) into (45.5)

$$\begin{aligned} C_A \, (\frac{3}{2} - \sqrt{3})\alpha + C_B \, (\frac{3}{2} - \frac{\sqrt{3}}{2})\alpha = 0 \\ C_A \, (\frac{3}{2} - \frac{\sqrt{3}}{2})\alpha + C_B \sqrt{3}\,\alpha = 0 \end{aligned} \qquad (45.10)$$

The coefficients C_A and C_B can be computed in the same way as in the HMO method. The difference is in the normalization condition, because the overlap integral S_{AB} is not equal to zero.

$$\Psi = C_A \Psi_A + C_B \Psi_B$$

$$\int \Psi^2 \, d\tau = 1 \qquad (45.11)$$

$$\int \Psi^2 \, d\tau = C_A^2 \int \Psi_A^2 \, d\tau + 2\,C_A C_B \int \Psi_A \Psi_B \, d\tau + C_B^2 \int \Psi_B^2 \, d\tau \qquad (45.12)$$

By comparing (45.2), (45.3), (45.4) with (45.11), (45.12), we find

$$1 = c_A^2 + c_B^2 + c_A c_B \qquad (45.13)$$

From (45.13) and (45.10) we obtain

$$c_A = 0.82$$
$$c_B = 0.30$$
$$\Psi = 0.82 \, \Psi_A + 0.30 \, \Psi_B$$

46.

$$\Psi = c_A \, \Psi_A + c_B \, \Psi_B + c_C \, \Psi_C + c_D \, \Psi_D + c_E \, \Psi_E \qquad (46.1)$$

$\underline{H_{AA}, \quad S_{AA}}$

Cycles $(\frac{a}{b})$, $(\frac{c}{d})$, $(\frac{e}{f})$ \qquad $x = 3$, $y = 3$

$$H_{AA} = \frac{2^3}{2^3} \left[Q + \frac{3}{2} \left\{ (ab) + (cd) + (ef) - 0 \right\} - \right.$$
$$\left. - \frac{1}{2} \left\{ (ab) + (af) + (bc) + (cd) + (de) + (ef) \right\} \right]$$

$$H_{AA} = Q + \frac{3}{2} \alpha \qquad\qquad S_{AA} = 1 \qquad (46.2)$$

$H_{BE}, \quad S_{BE}$

Cycles $(\frac{ac}{bf})$, $(\frac{e}{d})$ $x = 2, \quad y = 3$

$$H_{BE} = \frac{2^2}{2^3} \left[Q + \frac{3}{2} \left\{ (ab) + (af) + (bc) + (ed) \right\} - \right.$$

$$\left. - \frac{1}{2} \left\{ (ab) + (af) + (bc) + (cd) + (de) + (ef) \right\} \right]$$

$$H_{BE} = \frac{1}{2} Q + \frac{3}{2} \alpha \qquad S_{BE} = \frac{1}{2}$$

The remaining H_{KL} and S_{KL} can be computed in a similar way and substituted into Equation (45) of the Appendix.

$$c_A \left[(Q - E) + \frac{3}{2}\alpha \right] + c_B \left[\frac{1}{4} (Q - E) + \frac{3}{2}\alpha \right] + c_C \left[\frac{1}{2} (Q - E) + \frac{3}{2}\alpha \right] +$$

$$+ c_D \left[\frac{1}{2} (Q - E) + \frac{3}{2}\alpha \right] + c_E \left[\frac{1}{2} (Q - E) + \frac{3}{2}\alpha \right] = 0$$

$$c_A \left[\frac{1}{4} (Q - E) + \frac{3}{2}\alpha \right] + c_B \left[(Q - E) + \frac{3}{2}\alpha \right] + c_C \left[\frac{1}{2} (Q - E) + \frac{3}{2}\alpha \right] +$$

$$+ c_D \left[\frac{1}{2}(Q - E) + \frac{3}{2}\alpha \right] + c_E \left[\frac{1}{2} (Q - E) + \frac{3}{2}\alpha \right] = 0$$

$$c_A \left[\frac{1}{2} (Q - E) + \frac{3}{2}\alpha \right] + c_B \left[\frac{1}{2} (Q - E) + \frac{3}{2}\alpha \right] + c_C \left[(Q - E) \right] +$$

$$+ c_D \left[\frac{1}{4} (Q - E) + \frac{3}{2}\alpha \right] + c_E \left[\frac{1}{4} (Q - E) + \frac{3}{2}\alpha \right] = 0$$

$$c_A \left[\frac{1}{2} (Q - E) + \frac{3}{2}\alpha \right] + c_B \left[\frac{1}{2} (Q - E) + \frac{3}{2}\alpha \right] + c_C \left[\frac{1}{4} (Q - E) + \frac{3}{2}\alpha \right] + c_D \left[(Q - E) \right] + c_E \left[\frac{1}{4} (Q - E) + \frac{3}{2}\alpha \right] = 0$$

$$c_A \left[\frac{1}{2} (Q - E) + \frac{3}{2}\alpha \right] + c_B \left[\frac{1}{2} (Q - E) + \frac{3}{2}\alpha \right] + c_C \left[\frac{1}{4} (Q - E) + \frac{3}{2}\alpha \right] + c_D \left[\frac{1}{4} (Q - E) + \frac{3}{2}\alpha \right] + c_E \left[(Q - E) \right] = 0 \qquad (46.3)$$

Secular determinant:

$$
\begin{vmatrix}
(Q-E)+\frac{3}{2}\alpha & \frac{1}{4}(Q-E)+\frac{3}{2}\alpha & \frac{1}{2}(Q-E)+\frac{3}{2}\alpha & \frac{1}{2}(Q-E)+\frac{3}{2}\alpha & \frac{1}{2}(Q-E)+\frac{3}{2}\alpha \\
\frac{1}{4}(Q-E)+\frac{3}{2}\alpha & (Q-E)+\frac{3}{2}\alpha & \frac{1}{2}(Q-E)+\frac{3}{2}\alpha & \frac{1}{2}(Q-E)+\frac{3}{2}\alpha & \frac{1}{2}(Q-E)+\frac{3}{2}\alpha \\
\frac{1}{2}(Q-E)+\frac{3}{2}\alpha & \frac{1}{2}(Q-E)+\frac{3}{2}\alpha & (Q-E) & \frac{1}{4}(Q-E)+\frac{3}{2}\alpha & \frac{1}{4}(Q-E)+\frac{3}{2}\alpha \\
\frac{1}{2}(Q-E)+\frac{3}{2}\alpha & \frac{1}{2}(Q-E)+\frac{3}{2}\alpha & \frac{1}{4}(Q-E)+\frac{3}{2}\alpha & (Q-E) & \frac{1}{4}(Q-E)+\frac{3}{2}\alpha \\
\frac{1}{2}(Q-E)+\frac{3}{2}\alpha & \frac{1}{2}(Q-E)+\frac{3}{2}\alpha & \frac{1}{4}(Q-E)+\frac{3}{2}\alpha & \frac{1}{4}(Q-E)+\frac{3}{2}\alpha & (Q-E)
\end{vmatrix} = 0
$$

$$(46.4)$$

The five solutions for $(Q-E)$ then are: 2α; 2α; 0; $(1-\sqrt{13})\alpha$; $(1+\sqrt{13})\alpha$.

For the ground state:

$$
E = Q + (\sqrt{13} - 1)\alpha
$$

$$
E = Q + 2.6055\alpha
$$

$$(46.5)$$

From symmetry considerations, the structures A and B are equivalent as are C, D, and E; therefore

$$
C_A = C_B; \quad C_C = C_D = C_E
$$

$$(46.6)$$

Equation (46.3) simplifies to

$$
C_A \left[\frac{5}{2}(Q - E) + 6\alpha \right] + C_C \left[3(Q - E) + 9\alpha \right] = 0
$$

$$
C_A \left[3(Q - E) + 9\alpha \right] + C_C \left[\frac{9}{2}(Q - E) + 9\alpha \right] = 0
$$

$$(46.7)$$

Secular determinant:

$$
\begin{vmatrix}
\frac{5}{2}(Q - E) + 6\alpha & 3(Q - E) + 9\alpha \\
3(Q - E) + 9\alpha & \frac{9}{2}(Q - E) + 9\alpha
\end{vmatrix} = 0
$$

$$(46.8)$$

Solution with the lower energy:

$$
E = Q + (\sqrt{13} - 1)\alpha
$$

$$(46.5)$$

Computation of the resonance energy (difference between the ground state energy and the energy of the structure A):

$$E_A = \frac{\int \Psi_A \, H \, \Psi_A \, d\tau}{\int \Psi_A \, \Psi_A \, d\tau} = \frac{H_{AA}}{S_{AA}} = Q + 1.5\alpha$$

$$E - E_A = Q + (\sqrt{13} - 1)\alpha - Q - 1.5\alpha = 1.1055\alpha$$

Computation of coefficients:

We substitute (46.5) into (46.7) and divide by C_C, so that

$$\frac{C_A}{C_C} = C_A' \, , \quad \frac{C_C}{C_C} = C_C' = 1, \quad C_A' = \frac{1}{0.4341} \tag{46.9}$$

Then

$$\int \Psi^2 \, d\tau = \int (C_A \Psi_A + C_A \Psi_B + C_C \Psi_C + C_C \Psi_D + C_C \Psi_E)^2 \, d\tau$$

Normalization condition:

$$1 = 2 \, c_A^2 + 3 \, c_C^2 + 2 \, S_{AB} \, c_A^2 + 2 \, C_A C_C S_{AC} + 2 \, C_A C_C S_{AD} + \cdots \tag{46.10}$$

From (46.9) and (46.10) it follows that

$$C_A = 0.37; \quad C_C = 0.16$$

$$\Psi = 0.37 \, \Psi_A + 0.37 \, \Psi_B + 0.16 \, \Psi_C + 0.16 \, \Psi_D + 0.16 \, \Psi_E$$

47.

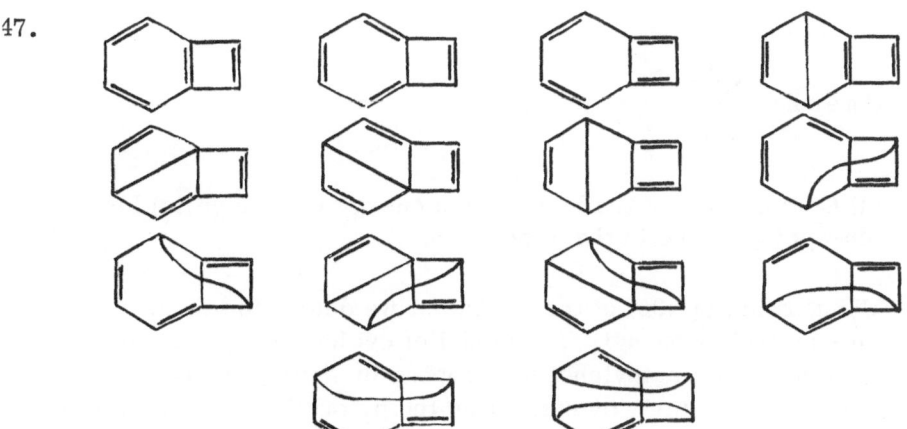

48. Benzene   5
     Naphthalene   42
     Anthracene   429
     Tetracene   4862

49.

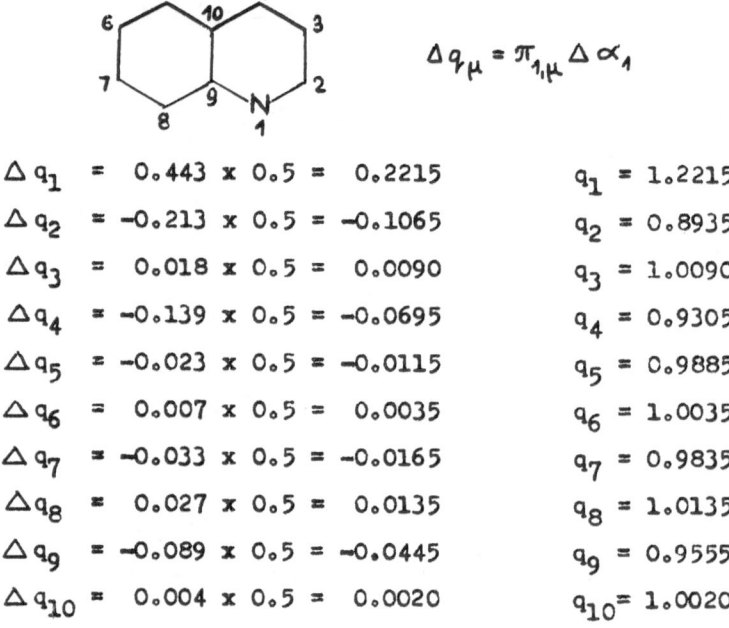

$$\Delta q_\mu = \pi_{1,\mu} \Delta \alpha_1$$

Δq_1 = 0.443 x 0.5 = 0.2215		q_1 = 1.2215
Δq_2 = -0.213 x 0.5 = -0.1065		q_2 = 0.8935
Δq_3 = 0.018 x 0.5 = 0.0090		q_3 = 1.0090
Δq_4 = -0.139 x 0.5 = -0.0695		q_4 = 0.9305
Δq_5 = -0.023 x 0.5 = -0.0115		q_5 = 0.9885
Δq_6 = 0.007 x 0.5 = 0.0035		q_6 = 1.0035
Δq_7 = -0.033 x 0.5 = -0.0165		q_7 = 0.9835
Δq_8 = 0.027 x 0.5 = 0.0135		q_8 = 1.0135
Δq_9 = -0.089 x 0.5 = -0.0445		q_9 = 0.9555
Δq_{10} = 0.004 x 0.5 = 0.0020		q_{10} = 1.0020

Check: $\displaystyle\sum_{\mu=1}^{10} q_\mu = 10.001$

50. (a) Computation of the localization energy (nucleophilic) for methy-
     lenecyclopropene in the μ-position.

     Using Equation (36) of the Appendix, we determine the orbital ener-
     gies of cyclopropenyl. The model of cyclopropenyl can be con-
     structed from methylenecyclopropene by setting the Coulomb in-
     tegral $\alpha_\mu = \infty$ (i.e., by extracting the p_z orbit of atom μ from the
     conjugation).

For $\alpha_\mu = \infty$ the right-hand side of Equation (36) is equal to

$$P = \frac{E - \delta_\nu}{\varrho_{\mu\nu}^2 + (E - \delta_\nu)\delta_\mu}$$

$$P = 0$$

The problem is solved graphically. We substitute into the left-hand side of Equation (36) for the expansion coefficients and the orbital energies of methylenecyclopropene

$$f(E) = \sum_i \frac{c_{i\mu}^2}{E - k_i}$$

$$f(E) = \frac{(0.2818)^2}{E - 2.170} + \frac{(-0.8152)^2}{E - 0.311} + \frac{0.0000}{E + 1.000} + \frac{(-0.5059)^2}{E + 1.481}$$

This function is then tabulated:

E	f(E)	E	f(E)	E	f(E)
2.5	0.609	0.8	1.413	-1.0	0.000
2.4	0.729	0.7	1.772	-1.1	0.176
2.3	1.013	0.6	2.372	-1.2	0.447
2.2	3.069	0.5	3.598	-1.3	0.978
2.15	-3.538	0.4	7.559	-1.4	2.748
2.1	-0.691	0.3	-60.317	-1.45	7.855
2.0	0.000	0.2	-5.876	-1.5	-13.857
1.9	0.200	0.1	-3.026	-1.6	-2.519
1.8	0.310	0.0	-2.001	-1.7	-1.520
1.7	0.390	-0.1	-1.467	-1.8	-1.137
1.6	0.459	-0.2	-1.134	-1.9	-0.931
1.5	0.526	-0.3	-0.903	-2.0	-0.799
1.4	0.596	-0.4	-0.729	-2.1	-0.708
1.3	0.673	-0.5	-0.588	-2.2	-0.639
1.2	0.761	-0.6	-0.468	-2.3	-0.584
1.1	0.867	-0.7	-0.357	-2.4	-0.541
1.0	1.072	-0.8	-0.249	-2.5	-0.505
0.9	1.173	-0.9	-0.134		

The function f(E) can be illustrated graphically (see below) by a discontinuous curve

$$f_1(E) = \frac{(0.2818)^2}{E - 2.170} + \frac{(-0.8152)^2}{E - 0.311} + \frac{(-0.5059)^2}{E + 1.481}$$

and by a straight line

$$f_2(E) = \frac{0}{E + 1.000}$$

(which is parallel to the y-axis and passes through $E = -1$).

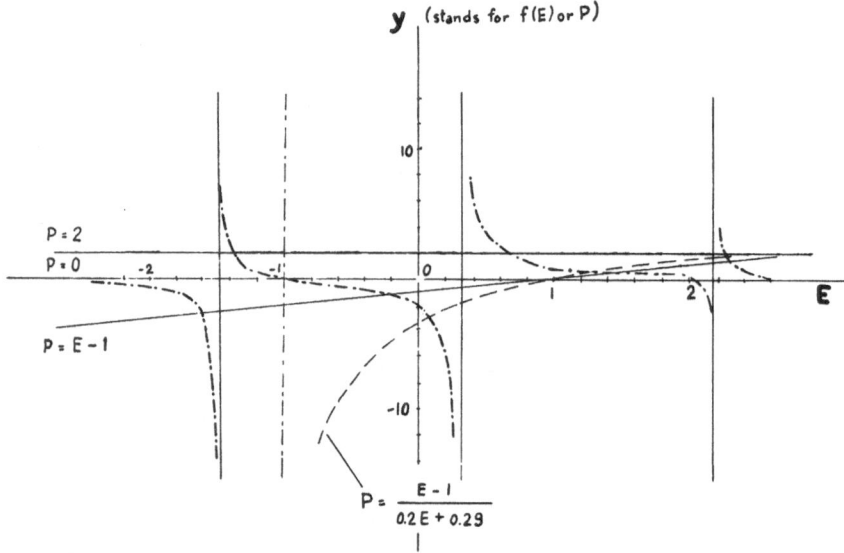

The intersections of functions f_1 and f_2 with the straight line $P = 0$ represent the required orbital energies of cyclopropenyl $(2; -1; -1)$.

The nucleophilic localization energy is

$$A_n = 4\alpha + [2(2.170 + 0.311)]\beta - 4\alpha - [(2 \times 2) - (2 \times 1)]\beta$$

$$A_n = 2.962\,\beta$$

Note: In this case, naturally enough, it would be easier to compute A_n by using the total π-electron energy of the anion of cyclopropenyl ⬡

(b) Computation of the orbital energies for model II:

$$P = \frac{E - 0}{0 + (E - 0) \, 0.5}$$

$$P = 2$$

The orbital energies can be determined from the intersections of the straight line $P = 2$ with the function $f(E)$:

$$E = \alpha + k\beta$$

$$k = 2.23; \ 0.66; \ -1.00; \ -1.37$$

Check (the sum of the eigenvalues is equal to the matrix trace):

$$2.23 + 0.66 - 1.00 - 1.37 \overset{\cdot}{=} 0.5$$

Computation of orbital energies for model III:

$$P = \frac{E - 1}{1 + (E - 1) \, 0}$$

$$P = E - 1$$

From the intersections of the straight line $P = E - 1$ with the function $f(E)$ it follows that

$$k = 2.28; \ 1.49; \ -0.16; \ -1.00; \ -1.60$$

Check:

$$2.28 + 1.49 - 0.16 - 1.00 - 1.60 \overset{\cdot}{=} 1.00$$

Computation of the orbital energies for model IV:

$$P = \frac{E - 1}{(0.7)^2 + (E - 1) \, 0.2}$$

$$P = \frac{E - 1}{0.2 \, E + 0.29}$$

This hyperbola is now tabulated

E	P	E	P
∞	5.00	0.8	−0.44
2.6	1.98	0.6	−0.98
2.5	1.90	0.4	−1.62
2.4	1.82	0.2	−2.42
2.2	1.64	0.0	−3.45
2.0	1.45	−0.2	−4.80
1.8	1.23	−0.4	−6.66
1.6	C.88	−0.6	−9.40
1.4	0.70	−0.8	−13.85
1.2	0.38	−1.45	∞
1.0	0.00	−2.4	+17.90

From the intersections of the hyperbola $P = \dfrac{E - 1}{0.2\,E + 0.29}$ with the function f(E) the orbital energies are determined:

$$k = 2.25;\ 1.31;\ 0.09;\ -1.00;\ -1.48$$

The last intersection (k = −1.48) is determined approximately for the second branch of the hyperbola.

Check:

$$2.25 + 1.31 + 0.09 - 1.00 - 1.48 \doteq 1.2$$

51.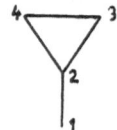

atom − atom

	1	2	3
1	0.402	−0.246	−0.078
2	−0.246	0.311	−0.032
3	−0.078	−0.032	0.434
4	−0.078	−0.032	−0.324

bond – bond

	1-2	2-3	3-4
1-2	0.265	-0.201	0.137
2-3	-0.201	0.329	-0.123
3-4	0.137	-0.123	0.110
2-4	-0.201	-0.005	-0.123

atom – bond

	1	2	3
1-2	-0.429	0.155	0.137
2-3	0.292	-0.045	-0.064
3-4	-0.155	-0.064	0.110
2-4	0.292	-0.045	-0.182

The polarizabilities can be computed by means of Equations (14), (15) and (16) of the Appendix.

For example, the atom – atom polarizability $\pi_{1,1}$:

$$\pi_{1,1} = 4 \left[\frac{(0.2818)^2 (0.0000)^2}{2.1700 + 1.0000} + \frac{(0.2818)^2 (0.5059)^2}{2.1700 + 1.4812} + \right.$$

$$\left. + \frac{(0.8152)^2 (0.0000)^2}{0.3111 + 1.0000} + \frac{(0.8152)^2 (0.5059)^2}{0.3111 + 1.4812} \right] =$$

$$= 4 \left[\frac{0.0794 \times 0.2559}{3.6512} + \frac{0.6645 \times 0.2559}{1.7923} \right] =$$

$$= 4 \left[\frac{0.020318}{3.6512} + \frac{0.17004}{1.7923} \right] = 4 \, (0.00556 + 0.0948) =$$

$$= 0.4015$$

The atom – bond polarizability $\pi_{1,34}$:

$$\pi_{1,34} = 4 \left[\frac{(0.2818)(0.0000) \, \{(0.5227)(-0.7071)+(0.5227)(0.7071)\}}{2.1700 + 1.0000} + \right.$$

$$\left. + \frac{(0.2818)(-0.5059) \, \{(0.5227)(-0.3020)+(0.5227)(-0.3020)\}}{2.1700 + 1.4812} + \right.$$

$$+ \frac{(-0.8152)(0.0000)\ \{(0.3682)(-0.7071)+(0.3682)(-0.7071)\}}{0.3111 + 1.0000} +$$

$$+ \frac{(-0.8152)(-0.5059)\ \{(0.3682)(-0.3020)+(0.3682)(-0.3020)\}}{0.3111 + 1.4812} \Bigg]$$

$$= 4 \left[\frac{(-0.1426)(-2 \times 0.1578)}{3.6512} + \frac{(0.4124)(-2 \times 0.1112)}{1.7923} \right] =$$

$$= 8 \left[\frac{0.0225}{3.6512} - \frac{0.0458}{1.7923} \right] = 8\ (0.006162 - 0.02555) = -0.155$$

The bond – bond polarizability $\pi_{12,23}$:

$$\pi_{12,23} = 2 \left[\frac{\{(0.6116)(-0.7071)+(0.5227)(0.0000)\}\ \{(0.2818)(0.0000)+(0.6116)(0.0000)\}}{2.1700 + 1.0000} + \right.$$

$$+ \frac{\{(0.6116)(-0.3020)+(0.5227)(0.7494)\}\ \{(0.2818)(0.7494)+(0.6116)(-0.5059)\}}{2.1700 + 1.4812} +$$

$$+ \frac{\{(-0.2536)(-0.7071)+(0.3682)(0.0000)\}\ \{(-0.8152)(0.0000)+(-0.2536)(0.0000)\}}{0.3111 + 1.0000} +$$

$$\left. + \frac{\{(-0.2536)(-0.3020)+(0.3682)(0.7494)\}\ \{(-0.8152)(0.7494)+(-0.2536)(-0.5059)\}}{0.3111 + 1.4812} \right] =$$

$$= 2 \left[\frac{(0.2070)(-0.0982)}{3.6512} + \frac{(0.3525)(-0.4826)}{1.7923} \right] = 2 \left[-\frac{0.02033}{3.6512} - \frac{0.1701}{1.7923} \right] =$$

$$= 2\ (-0.005568 - 0.0949) = -0.201$$

Note: If we use coefficients which are given to only three or four decimal places, we obtain only approximate agreement with the more accurate polarizabilities mentioned above.

Check:

$$\sum_{\nu=1}^{4} \pi_{\mu\nu} = 0; \qquad \sum_{\varrho\sigma} \pi_{\mu\nu,\varrho\sigma} = 0; \qquad \sum_{\varrho\sigma} \pi_{\mu,\varrho\sigma} = 0$$

Atom – atom polarizability for position 1:

$$0.402 - 0.246 - 0.078 - 0.078 = 0$$

Atom – bond polarizability for position 1:

$$0.429 - 0.292 + 1.155 - 0.292 = 0$$

52. The π-electron densities of cations and anions of odd alternant hydrocarbons are defined by the coefficients of the nonbonding molecular orbital (NBMO, k = 0). According to Longuet-Higgins[16] these coefficients can be determined without solving the secular equation. $2n-1$ conjugate atoms can be divided into n starred atoms and $n-1$ unstarred atoms. Then the sum of NBMO coefficients around the unstarred position is equal to zero. Let the value of

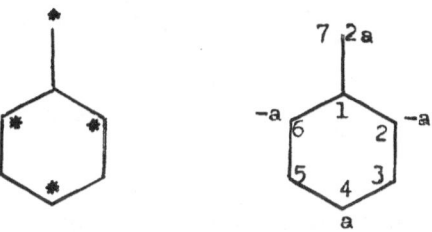

the coefficient in position 4 for benzyl be a. Since the sums of the coefficients around positions 1, 3, and 5 must be zero, the values of the coefficients in the remaining (starred) positions are uniquely determined. The NBMO must satisfy the normal-ization condition, i.e.,

$$(-a)^2 + (a)^2 + (-a)^2 + (2a)^2 = 1$$
$$7a^2 = 1$$
$$a = \frac{1}{\sqrt{7}} = 0.378$$

Therefore, the NBMO can be written as

$$\varphi = -0.378\, \chi_2 + 0.378\, \chi_4 - 0.378\, \chi_6 + 0.756\, \chi_7$$

Odd alternant hydrocarbons have unit π-electron densities in all positions and symmetrically arranged orbital levels about the NBMO. In the case of the benzyl system the situation is as follows:

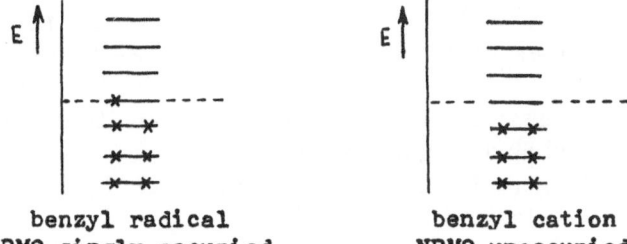

benzyl radical
NBMO singly occupied

benzyl cation
NBMO unoccupied

The benzyl radical has unit π-electron densities. The squares of the NBMO coefficients represent deviations from the unit π-electron distribution in the benzyl cation.

$$q_1 = 1$$
$$q_2 = 1 - (-0.378)^2 = 0.857$$
$$q_3 = 1$$
$$q_4 = 1 - (0.378)^2 = 0.857$$
$$q_5 = 1$$
$$q_6 = 1 - (-0.378)^2 = 0.857$$
$$q_7 = 1 - (2 \times 0.378)^2 = 0.428$$

C h e c k (sum of π-electron densities equals the number of π-electrons):

$$1 + 0.857 + 1 + 0.857 + 1 + 0.857 + 0.428 = 5.999$$

N o t e :   For the anion the absolute values of the deviations from the unit π-electron densities are the same as for the cation.

53. Let us consider a limiting case in which the electronegativity of nitrogen is much larger than that of the sp^2 hybridized carbon atom. In this case two of the fourteen conjugate π-electrons will be localized on nitrogen and the π-electron distribution created on the torsos can be represented by the Longuet-Higgins model:

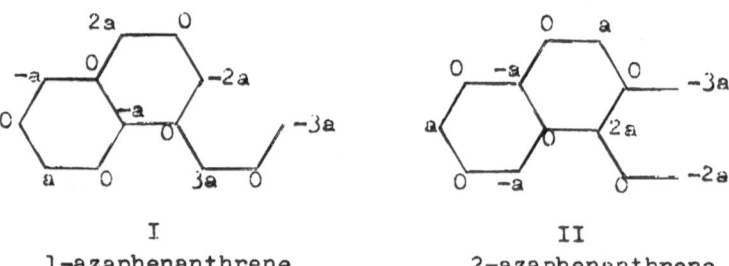

I   II

1-azaphenanthrene   2-azaphenanthrene

I: Largest decrease of π-electron density in positions 2 and 4 (and next largest in position 9).

II: Largest decrease of π-electron density in position 1 (and next largest in position 3).

64. Use Equations (21) and (23) of the Appendix:

$$\Delta q_\mu = \pi_{\mu\nu} \Delta \alpha_\nu$$

$$\Delta p_{\mu\nu} = \pi_{\mu\nu,\rho\sigma} \Delta \beta_{\rho\sigma}$$

Thiophene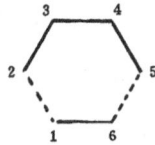

$$\Delta p_{12} = \pi_{12,12} \Delta\beta_{12} + \pi_{12,56} \Delta\beta_{56} = (0.241)(-0.4) +$$
$$+ (0.130)(-0.4) = -0.148$$

$$\Delta p_{23} = \pi_{23,12} \Delta\beta_{12} + \pi_{23,56} \Delta\beta_{12} = (-0.204)(-0.4) +$$
$$+ (-0.092)(-0.4) = 0.118$$

$$\Delta p_{34} = \pi_{34,12} \Delta\beta_{12} + \pi_{34,56} \Delta\beta_{56} = 2(0.130)(-0.4) = -0.104$$

$$p_{12} = 0.666 - 0.148 = 0.518$$
$$p_{23} = 0.666 + 0.118 = 0.784$$
$$p_{34} = 0.666 - 0.104 = 0.562$$

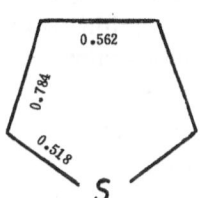

Pyridine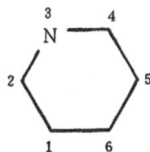

$$\Delta q_1 = \pi_{1,3} \Delta\alpha_3 = 0.00925 \times 0.5 = 0.005$$
$$\Delta q_2 = \pi_{2,3} \Delta\alpha_3 = (-0.157) \times 0.5 = -0.078$$
$$\Delta q_3 = \pi_{3,3} \Delta\alpha_3 = 0.398 \times 0.5 = 0.199$$
$$\Delta q_6 = \pi_{6,3} \Delta\alpha_3 = (-0.102) \times 0.5 = -0.051$$

$q_1 = 1.000 + 0.005 = 1.005$

$q_2 = 1.000 - 0.078 = 0.922$

$q_3 = 1.000 + 0.199 = 1.199$

$q_6 = 1.000 - 0.051 = 0.949$

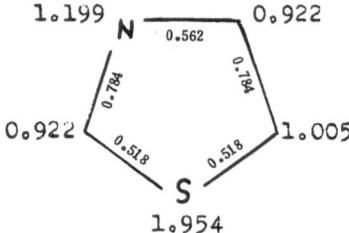

1,3-Thiazol (by combining the two molecular diagrams)

The π-electron density on sulfur is represented by the sum q_1 and q_6.

By comparing this molecular diagram with that one obtained by solving the secular equations one sees clearly that the perturbation method yields two positions with equally low π-electron density. Thus it does not succeed in indicating the center of lowest π-electron density.

55. Use Equation (39) of the Appendix:

$$\alpha_\mu = \alpha_0 + (1 - q_\mu)\,\omega\,\beta_0 \qquad\qquad \begin{aligned} q_1 &= 0.5 \\ q_2 &= 1 \end{aligned}$$

$$\alpha_1 = \alpha_0 + (1 - 0.5)\,1.4\,\beta_0 = \alpha_0 + 0.7\,\beta_0$$

$$\alpha_2 = \alpha_0$$

Sx: $(-k + 0.7)c_1 + c_2 = 0$

$-kc_2 + 2c_1 = 0$

$$\begin{vmatrix} -k + 0.7 & 1 \\ 2 & -k \end{vmatrix} = 0 \qquad\qquad \begin{aligned} k_1 &= 1.805 \\ k_3 &= -1.105 \end{aligned}$$

Ax: $(-k + 0.7)c_1 = 0$ $k_2 = 0.700$

$(-1.805 + 0.7)c_1 + c_2 = 0$

$-1.805c_2 + 2c_1 = 0$

$2c_1^2 + c_2^2 = 0$ $q_1 = 0.621$
                                           $q_2 = 0.757$

$\alpha_1 = \alpha_0 + (1 - 0.621)\ 1.4\ \beta_0 = \alpha_0 + 0.530\ \beta_0$

$\alpha_2 = \alpha_0 + (1 - 0.757)\ 1.4\ \beta_0 = \alpha_0 + 0.340\ \beta_0$

Sx: $(-k + 0.530)c_1 + c_2 = 0$

$(-k + 0.340)c_2 + 2c_1 = 0$

$$\begin{vmatrix} -k + 0.530 & 1 \\ 2 & -k + 0.340 \end{vmatrix} = 0 \qquad \begin{aligned} k_1 &= 1.852 \\ k_3 &= -0.982 \end{aligned}$$

Ax: $(-k + 0.530)c_1 = 0$ $k_2 = 0.530$

$q_1 = 0.534$ $q_2 = 0.934$

etc.

Number of steps	q_1	q_2
0	0.500	1.000
1	0.621	0.757
2	0.534	0.934
3	0.597	0.806
4	0.552	0.896
5	0.584	0.830
6	0.560	0.880
∞	0.571	0.858

56. Example of the computation of the matrix elements according to Janssen and Sandström [Tetrahedron 20:2339 (1964)]:

diagonal elements $H_{\mu\mu} = \delta_\mu + \omega (n_\mu - q_\mu)$

nondiagonal elements $H_{\mu\nu} = \rho_{\mu\nu} (1 + 0.5\ p_{\mu\nu})$

For acrolein:

$$H_{11} = 0 + 1(1 - 0.813) = 0.187$$
$$H_{22} = 0 + 1(1 - 1.035) = -0.035$$
$$H_{33} = 0 + 1(1 - 0.699) = 0.301$$
$$H_{44} = 1 + 1(1 - 1.453) = 0.547$$
$$H_{12} = 1(1 + 0.5 \times 0.886) = 1.443$$
$$H_{23} = 1(1 + 0.5 \times 0.453) = 1.226$$
$$H_{34} = 1.2(1 + 0.5 \times 0.810) = 1.686$$

Matrix:

$$\begin{pmatrix} 0.187 & 1.443 & 0 & 0 \\ 1.443 & -0.035 & 1.226 & 0 \\ 0 & 1.226 & 0.301 & 1.686 \\ 0 & 0 & 1.686 & 0.547 \end{pmatrix}$$

By diagonalization of this matrix, new eigenvalues and expansion coefficients are obtained; they can be used to compute the π-electron densities and bond orders, which in turn are utilized to compute the new matrix elements. This procedure is repeated until self-consistency is reached.

For urea the matrix elements are computed in a similar way (beware: $n_N = 2$). With the help of the HMO data the following matrix is obtained in the first step:

$$\begin{pmatrix} 1.704 & 1.502 & 0 & 0 \\ 1.502 & 0.298 & 1.502 & 1.570 \\ 0 & 1.502 & 1.704 & 0 \\ 0 & 1.570 & 0 & 1.293 \end{pmatrix}$$

The procedure is repeated until self-consistency is reached. The results of the HMO and SC computations are given in the table on p. 23.

From a comparison of SC and HMO(B) π-electron densities and bond orders one can see that a simple HMO computation with a suitable choice of parameters yields data very near to the SC results.

57. The matrix elements can be calculated using Equations (40) and (41) of the Appendix:

$$H_{\mu\mu} = \delta_\mu$$

$$H_{\mu\nu} = \exp\left[-2.683(0.120 - 0.180\ p_{\mu\nu})\right]$$

$$H_{12} = \exp\left[-2.683(0.120 - 0.180 \times 0.758)\right] =$$

$$= \exp\left[0.0429\right] = 1.044$$

$$H_{23} = \exp\left[-2.683(0.120 - 0.180 \times 0.453)\right] =$$

$$= \exp\left[-0.102\right] = 0.903$$

$$H_{34} = \exp\left[-2.683(0.120 - 0.180 \times 0.818)\right] =$$

$$= \exp\left[0.0724\right] = 1.075$$

$$\begin{pmatrix} 0 & 1.044 & 0 & 0 \\ 1.044 & 0 & 0.903 & 0.903 \\ 0 & 0.903 & 0 & 1.075 \\ 0 & 0.903 & 1.075 & 0 \end{pmatrix}$$

The solution of the appropriate secular equations yields the new bond orders, from which the new matrix elements can be computed; this is repeated until self-consistency is reached.

58.

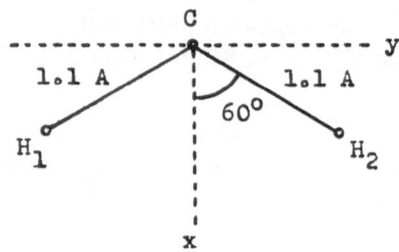

	$1s(H_1)$	$2s(C)$	$2p_x$	$2p_y$	$2p_z$	$1s(H_2)$
$1s(H_1)$	1	0.56	-0.23	-0.40	0	0.24
$2s(C)$		1	0	0	0	0.56
$2p_x$			1	0	0	-0.23
$2p_y$				1	0	0.40
$2p_z$					1	0
$1s(H_2)$						1

$S_{\mu\nu}$

	$1s(H_1)$	$2s(C)$	$2p_x$	$2p_y$	$2p_z$	$1s(H_2)$
$1s(H_1)$	-13.6	-17.15	5.03	8.75	0	-5.71
$2s(C)$		-21.4	0	0	0	-17.15
$2p_x$			-11.4	0	0	5.03
$2p_y$				-11.4	0	-8.75
$2p_z$					-11.4	0
$1s(H_2)$						-13.6

$H_{\mu\nu}$

59. (a)

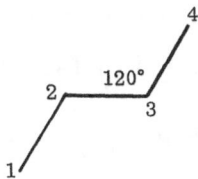

The interatomic distances r can be determined from trigonometrical formulas.

$$r_{\mu\nu} \; (\mathring{A})$$

	1	2	3	4
1	—	1.40000	2.42487	3.70405
2		—	1.40000	2.42487
3			—	1.40000
4				—

According to Mataga and Nishimoto the repulsion integrals have the form

$$\tilde{\gamma}_{\mu\nu} = \frac{14.3986}{1.3674 + r_{\mu\nu}}$$

$$\tilde{\gamma}_{\mu\nu} \; (eV)$$

	1	2	3	4
1	10.53000	5.20293	3.79690	2.83915
2		10.53000	5.20293	3.79690
3			10.53000	5.20293
4				10.53000

The bond orders can be computed from the HMO expansion coefficients:

$$\tfrac{1}{2} P_{\mu\nu}$$

	1	2	3	4
1	0.50000[*]	0.44722	0.00000	−0.22360
2		0.50000	0.22360	0.00000
3			0.50000	0.44722
4				0.50000

---

[*]With alternant hydrocarbons all π-electron densities are equal to 1.

The F-matrix elements are defined as

$$F_{\mu\mu} = -11.22 + \frac{1}{2} P_{\mu\mu} \gamma_{\mu\mu} + \sum_{\sigma(\neq\mu)} (P_{\sigma\sigma} - 1) \gamma_{\mu\sigma}$$

$$F_{\mu\nu} = \beta_{\mu\nu} - \frac{1}{2} P_{\mu\nu} \gamma_{\mu\nu} = -2.318 - \frac{1}{2} P_{\mu\nu} \gamma_{\mu\nu}$$

F-matrix (in eV), first iteration

	1	2	3	4
1	-5.95500	-4.64485	0.00000	0.63483
2		-5.95500	-3.48138	0.00000
3			-5.95500	-4.64485
4				-5.95500

(b) If the computation is accurate to three decimal places, the F-matrix of the seventh iteration will be identical with the F-matrix of the eighth iteration. This means that self-consistency has been reached.

An edited output from a computer follows, in which one may see the whole SCF computation for transbutadien. Small numerical discrepancies between this and the calculation presented above are due to the fact that the computer calculation was performed with more valid decimal places (i.e., with greater accuracy). The SCF computation is concluded after k iterations, when the following holds for all orbital energies (ε) of the k-th iteration:

$$|\varepsilon_i (k) - \varepsilon_i (k - 1)| < 0.00001 \text{ eV}$$

(criterion of self-consistency).

GAMMA INTEGRALS

1	1	10.53000	1	2	5.20296	1	3	3.79684
2	1	5.20296	2	2	10.53000	2	3	5.20296
3	1	3.79684	3	2	5.20296	3	3	10.53000
4	1	2.83915	4	2	3.79684	4	3	5.20296
1	4	2.83915						
2	4	3.79684						
3	4	5.20296						
4	4	10.53000						

F MATRIX
(constructed from the HMO coefficients)

1	1	−5.95500	1	2	−4.64483	1	3	0.00000
2	1	−4.64483	2	2	−5.95500	2	3	−3.48142
3	1	0.00000	3	2	−3.48142	3	3	−5.95500
4	1	0.63485	4	2	0.00000	4	3	−4.64483

1	4	0.63485
2	4	0.00000
3	4	−4.64483
4	4	−5.95500

1 ITERATION

SCF ENERGIES AND COEFFICIENTS

	1	2	3	4	
	−12.4587	−9.61211	−2.29789	.548674	
1	0.38564	−0.59269	0.59269	−0.38564	1
2	0.59269	−0.38564	−0.38564	0.59269	2
3	0.59269	0.38564	−0.38564	−0.59269	3
4	0.38564	0.59269	0.59269	0.38564	4

CHARGE DENSITIES AND BOND ORDERS

1	1	1.00000	1	2	0.91427
2	2	1.00000	2	3	0.40511
3	3	1.00000	3	4	0.91427
4	4	1.00000			

THE F MATRIX AFTER 1 ITERATION

1	1	−5.95500	1	2	−4.69644	1	3	0.00000
2	1	−4.69644	2	2	−5.95500	2	3	−3.37189
3	1	0.00000	3	2	−3.37189	3	3	−5.95500
4	1	0.57509	4	2	−0.00000	4	3	−4.69644

1	4	0.57509
2	4	−0.00000
3	4	−4.69644
4	4	−5.95500

2 ITERATION

SCF ENERGIES AND COEFFICIENTS

	1	2	3	4	
	-12.4476	-9.65084	-2.25916	.537640	
1	0.39134	-0.58894	0.58894	-0.39134	1
2	0.58894	-0.39134	-0.39135	0.58894	2
3	0.58894	0.39134	-0.39134	-0.58894	3
4	0.39134	0.58894	0.58894	0.39135	4

CHARGE DENSITIES AND BOND ORDERS

1	1	1.00000	1	2	0.92191
2	2	1.00000	2	3	0.38740
3	3	1.00000	3	4	0.92191
4	4	1.00000			

THE F MATRIX AFTER 2 ITERATIONS

1	1	-5.95500	1	2	-4.71634	1	3	-0.00000
2	1	-4.71634	2	2	-5.95500	2	3	-3.32580
3	1	-0.00000	3	2	-3.32580	3	3	-5.95500
4	1	0.54994	4	2	0.00000	4	3	-4.71634

1	4	0.54994
2	4	0.00000
3	4	-4.71634
4	4	-5.95500

THE F MATRIX AFTER 10 ITERATIONS

1	1	-5.95500	1	2	-4.72971	1	3	0.00000
2	1	-4.72971	2	2	-5.95500	2	3	-3.29337
3	1	0.00000	3	2	-3.29337	3	3	-5.95500
4	1	0.53224	4	2	0.00000	4	3	-4.72971

1	4	0.53224
2	4	0.00000
3	4	-4.72971
4	4	-5.95500

11 ITERATION

SCF ENERGIES AND COEFFICIENTS

	1	2	3	4	
	−12.4374	−9.67630	−2.23370	.527425	
1	0.39531	−0.58629	0.58629	−0.39531	1
2	0.58629	−0.39531	−0.39531	0.58629	2
3	0.58629	0.39531	−0.39531	−0.58629	3
4	0.39531	0.58629	0.58629	0.39531	4

CHARGE DENSITIES AND BOND ORDERS

1	1	1.00000	1	2	0.92706
2	2	1.00000	2	3	0.37492
3	3	1.00000	3	4	0.92706
4	4	1.00000			

THE F MATRIX AFTER 11 ITERATIONS

1	1	−5.95500	1	2	−4.72972	1	3	0.00000
2	1	−4.72972	2	2	−5.95500	2	3	−3.29335
3	1	0.00000	3	2	−3.29335	3	3	−5.95500
4	1	0.53223	4	2	−0.00000	4	3	−4.72972

1	4	0.53223
2	4	−0.00000
3	4	−4.72972
4	4	−5.95500

12 ITERATIONS

SCF ENERGIES AND COEFFICIENTS

	1	2	3	4	
	−12.4374	−9.67630	−2.23370	.527423	
1	0.39531	−0.58628	0.58629	−0.39531	1
2	0.58629	−0.39531	−0.39531	0.58628	2
3	0.58628	0.39531	−0.39531	−0.58629	3
4	0.39531	0.58629	0.58628	0.39531	4

CHARGE DENSITIES AND BOND ORDERS

1	1	1.00000	1	2	0.92706
2	2	1.00000	2	3	0.37492
3	3	1.00000	3	4	0.92706
4	4	1.00000			

60. Use Equation (58) of the Appendix:

$$\mathcal{E}_\pi = \sum_\mu \left(-11.22 + \frac{1}{4}\gamma_{\mu\mu}\right) + 2 \sum_{\substack{\mu < \nu \\ (\text{adjacent})}} P_{\mu\nu}(-2.39) -$$

$$- \frac{1}{2} \sum_{\substack{\mu < \nu \\ (\text{also un-} \\ \text{adjacent})}} P^2_{\mu\nu}\gamma_{\mu\nu} - \sum_{\mu < \nu} \gamma_{\mu\nu}$$

Repulsion integrals $\gamma_{\mu\nu}$

	1	2	3	4	5	6	7	8	9	10
1	10.530	5.203	3.797	3.455	2.839	2.586	2.839	3.797	5.203	3.797
2		10.530	5.203	3.797	2.586	2.244	2.316	2.839	3.797	3.455
3			10.530	5.203	2.839	2.316	2.244	2.586	3.455	3.797
4				10.530	3.797	2.839	2.586	2.839	3.797	5.203
5					10.530	5.203	3.797	3.455	3.797	5.203
6						10.530	5.203	3.797	3.455	3.797
7							10.530	5.203	3.797	3.455
8								10.530	5.203	3.797
9									10.530	5.203
10										10.530

$$P^2_{\mu\nu}$$

	1	2	3	4	5	6	7	8	9	10
1	1.000	0.549	0.000	0.128	0.007	0.000	0.024	0.000	0.292	0.000
2		1.000	0.345	0.000	0.000	0.020	0.000	0.024	0.000	0.062
3			1.000	0.549	0.024	0.000	0.020	0.000	0.062	0.000
4				1.000	0.000	0.024	0.000	0.007	0.000	0.292
5					1.000	0.549	0.000	0.128	0.000	0.292
6						1.000	0.345	0.000	0.062	0.000
7							1.000	0.549	0.000	0.062
8								1.000	0.292	0.000
9									1.000	0.293
10										1.000

$$P^2_{\mu\nu} \; \gamma_{\mu\nu}$$

	1	2	3	4	5	6	7	8	9	10
1	–	2.856	0.000	0.442	0.020	0.000	0.068	0.000	1.519	0.000
2		–	1.795	0.000	0.000	0.045	0.000	0.068	0.000	0.214
3			–	2.856	0.068	0.000	0.045	0.000	0.214	0.000
4				–	0.000	0.068	0.000	0.020	0.000	1.519
5					–	2.856	0.000	0.442	0.000	1.519
6						–	1.795	0.000	0.214	0.000
7							–	2.856	0.000	0.214
8								–	1.519	0.000
9									–	1.524
10										–

$$\mathcal{E}_\pi = 10 \left(-11.22 + \tfrac{1}{4} 10.53\right) - 4.78 \underset{\substack{\mu < \nu \\ \text{(adjacent)}}}{\sum\sum} P_{\mu\nu} -$$

$$- \tfrac{1}{2} \left(P^2_{12}\, \gamma_{12} + P^2_{13}\, \gamma_{13} + \ldots + P^2_{110}\, \gamma_{110} + \right.$$

$$\left. + P^2_{23}\, \gamma_{23} + \ldots + P^2_{210}\, \gamma_{210} + \ldots \right) -$$

$$- \underset{\substack{\mu < \nu \\ \text{(also un-} \\ \text{adjacent)}}}{\sum\sum} \gamma_{\mu\nu} = -85.88 - 32.69 - 12.38 - 167.22 =$$

$$= -298.55 \text{ eV}$$

61. (a) The SCF orbital energy is given by

$$E_i = H^c_{ii} + \sum_j^{occ} (2 J_{ij} - K_{ij})$$

$$E_1 = H^c_{11} + 2 J_{11} - K_{11} + 2 J_{12} - K_{12} =$$

$$= -28.426 + 2 \times 6.304 - 6.304 + 2 \times 5.692 - 1.701$$

$$= -12.437 \text{ eV}$$

$E_2 = -9.676$ eV

$E_3 = -2.234$ eV

(b)
$$\varepsilon = 2 \sum_{i}^{occ} H_{ii}^c + \sum_{i}^{occ} \sum_{j}^{occ} (2 J_{ij} - K_{ij})$$

$\varepsilon = 2 H_{11}^c + 2 H_{22}^c + 2 J_{11} - K_{11} + 2 J_{12} - K_{12} +$

$\quad + 2 J_{21} - K_{21} + 2 J_{22} - K_{22} =$

$\quad = -56.852 - 50.444 + 6.304 + 2 \times 5.692 - 1.701 +$

$\quad + 2 \times 5.692 - 1.701 + 5.861 =$

$\quad = -75.762$ eV

The sum of orbital energies becomes

$$E' = (E_1 + E_2) \times 2 = -44.227 \text{ eV}$$

(c) The singlet – singlet transition energy is given by

$^1\Delta E_{2 \rightarrow 3} = E_3 - E_2 - J_{23} + 2 K_{23}$

$^1\Delta E_{2 \rightarrow 3} = -2.234 + 9.676 - 5.861 + 2 \times 1.775 =$

$\quad = 5.132$ eV

For the singlet – triplet transition energy we obtain

$^3\Delta E_{2 \rightarrow 3} = E_3 - E_2 - J_{23} = 1.582$ eV

The singlet – triplet splitting is

$^1\Delta E_{2 \rightarrow 3} - {}^3\Delta E_{2 \rightarrow 3} = E_3 - E_2 - J_{23} + 2 K_{23} - E_3 +$

$\quad + E_2 + J_{23} = 2 K_{23} =$

$\quad = 3.551$ eV

With $n \rightarrow \pi^*$ transitions the respective K_{ij} integrals calculated by the all-valence electrons methods are very little, because the n and π^* orbitals are orthogonal (within the ZDO approximation).

(d)

$$J_{11} = \iint \varphi_1(1)\, \varphi_1(2)\, \frac{e^2}{r_{12}}\, \varphi_1(1)\, \varphi_1(2)\, d\tau_1\, d\tau_2 =$$

$$= \sum_{\mu} \sum_{\nu} c_{1\mu}^2\, c_{1\nu}^2 \iint \chi_1(1)\, \chi_1(2)\, \frac{e^2}{r_{12}}\, \chi_1(1)\, \chi_1(2)\, d\tau_1\, d\tau_2 =$$

$$= \sum_{\mu} \sum_{\nu} c_{1\mu}^2\, c_{1\nu}^2\, \gamma_{\mu\nu}$$

$$J_{11} = c_{11}^2\, c_{11}^2 \gamma_{11} + c_{11}^2\, c_{12}^2 \gamma_{12} + c_{11}^2\, c_{13}^2 \gamma_{13} +$$
$$+ c_{11}^2\, c_{14}^2 \gamma_{14} + c_{12}^2\, c_{11}^2 \gamma_{21} + c_{12}^2\, c_{12}^2 \gamma_{22} +$$
$$+ c_{12}^2\, c_{13}^2 \gamma_{23} + c_{12}^2\, c_{14}^2 \gamma_{24} + c_{13}^2\, c_{11}^2 \gamma_{31} +$$
$$+ c_{13}^2\, c_{12}^2 \gamma_{32} + c_{13}^2\, c_{13}^2 \gamma_{33} + c_{13}^2\, c_{14}^2 \gamma_{34} +$$
$$+ c_{14}^2\, c_{11}^2 \gamma_{41} + c_{14}^2\, c_{12}^2 \gamma_{42} + c_{14}^2\, c_{13}^2 \gamma_{43} + \cdot$$
$$+ c_{14}^2\, c_{14}^2 \gamma_{44}$$

$$J_{11} = 0.257 + 0.279 + 0.204 + 0.069 + 0.279 + 1.244 +$$
$$+ 0.615 + 0.204 + 0.204 + 0.615 + 1.244 + 0.279 +$$
$$+ 0.069 + 0.204 + 0.279 + 0.257 = 6.304 \text{ eV}$$

$$J_{12} = \iint \varphi_1(1)\, \varphi_2(2)\, \frac{e^2}{r_{12}}\, \varphi_1(1)\, \varphi_2(2)\, d\tau_1\, d\tau_2 =$$

$$= \sum_{\mu} \sum_{\nu} c_{1\mu}^2\, c_{2\nu}^2 \iint \chi_1(1)\, \chi_2(2)\, \frac{e^2}{r_{12}}\, \chi_1(1)\, \chi_2(2)$$

$$d\tau_1\, d\tau_2 =$$

$$= \sum_{\mu} \sum_{\nu} c_{1\mu}^2\, c_{2\nu}^2\, \gamma_{\mu\nu}$$

$$J_{12} = c_{11}^2 \, c_{21}^2 \, \mathcal{J}_{11} + c_{11}^2 \, c_{22}^2 \, \mathcal{J}_{12} + c_{11}^2 \, c_{23}^2 \, \mathcal{J}_{13} + $$

$$+ \, c_{11}^2 \, c_{24}^2 \, \mathcal{J}_{14} + c_{12}^2 \, c_{21}^2 \, \mathcal{J}_{21} + c_{12}^2 \, c_{22}^2 \, \mathcal{J}_{22} + $$

$$+ \, c_{12}^2 \, c_{23}^2 \, \mathcal{J}_{23} + c_{12}^2 \, c_{24}^2 \, \mathcal{J}_{24} + c_{13}^2 \, c_{21}^2 \, \mathcal{J}_{31} + $$

$$+ \, c_{13}^2 \, c_{22}^2 \, \mathcal{J}_{32} + c_{13}^2 \, c_{23}^2 \, \mathcal{J}_{33} + c_{13}^2 \, c_{24}^2 \, \mathcal{J}_{34} + $$

$$+ \, c_{14}^2 \, c_{21}^2 \, \mathcal{J}_{41} + c_{14}^2 \, c_{22}^2 \, \mathcal{J}_{42} + c_{14}^2 \, c_{23}^2 \, \mathcal{J}_{43} + $$

$$+ \, c_{14}^2 \, c_{24}^2 \, \mathcal{J}_{44} = 5.693 \text{ eV}$$

$$K_{12} = \varphi_1(1) \, \varphi_2(2) \, \frac{e^2}{r_{12}} \, \varphi_2(1) \, \varphi_1(2) \, d\tau_1 \, d\tau_2 = $$

$$= \sum_\mu \sum_\nu c_{1\mu} \, c_{2\nu} \, c_{2\mu} \, c_{1\nu} \, \chi_1(1) \, \chi_2(2) \, \frac{e^2}{r_{12}} \, \chi_2(1) \, \chi_1(2) $$

$$d\tau_1 \, d\tau_2 = $$

$$= \sum_\mu \sum_\nu c_{1\mu} \, c_{2\mu} \, c_{1\nu} \, c_{2\nu} \, \mathcal{J}_{\mu\nu}$$

$$K_{12} = c_{11}^2 \, c_{21}^2 \, \mathcal{J}_{11} + c_{11} \, c_{21} \, c_{12} \, c_{22} \, \mathcal{J}_{12} + $$

$$+ \, c_{11} \, c_{21} \, c_{13} \, c_{23} \, \mathcal{J}_{13} + c_{11} \, c_{21} \, c_{14} \, c_{24} \, \mathcal{J}_{14} + $$

$$+ \, c_{12} \, c_{22} \, c_{11} \, c_{21} \, \mathcal{J}_{21} + c_{12}^2 \, c_{22}^2 \, \mathcal{J}_{22} + $$

$$+ \, c_{12} \, c_{22} \, c_{13} \, c_{23} \, \mathcal{J}_{23} + c_{12} \, c_{22} \, c_{14} \, c_{24} \, \mathcal{J}_{24} + $$

$$+ \, c_{13} \, c_{23} \, c_{11} \, c_{21} \, \mathcal{J}_{31} + c_{13} \, c_{23} \, c_{12} \, c_{22} \, \mathcal{J}_{32} + $$

$$+ \, c_{13}^2 \, c_{23}^2 \, \mathcal{J}_{33} + c_{13} \, c_{23} \, c_{14} \, c_{24} \, \mathcal{J}_{34} + $$

$$+ \, c_{14} \, c_{24} \, c_{11} \, c_{21} \, \mathcal{J}_{41} + c_{14} \, c_{24} \, c_{12} \, c_{22} \, \mathcal{J}_{42} + $$

$$+ \, c_{14} \, c_{24} \, c_{13} \, c_{23} \, \mathcal{J}_{43} + c_{14}^2 \, c_{24}^2 \, \mathcal{J}_{44} = $$

$$= 1.701 \text{ eV}$$

$$H_{ii}^c = \int \varphi_i(1)\, \hat{H}^c\, \varphi_i(1)\, d\tau =$$

$$= \sum_\mu \sum_\nu c_{i\mu}\, c_{i\nu} \int \chi_\mu \hat{H}^c \chi_\nu\, d\tau =$$

$$= \sum_\mu \sum_\nu c_{i\mu}\, c_{i\nu}\, H_{\mu\nu}^c$$

$$H_{\mu\mu}^c = -I_\mu - \sum_{(\nu \neq \mu)} \hat{\gamma}_{\mu\nu} \qquad\qquad I = 11.22 \text{ eV}$$

$$H_{\mu\nu}^c = -2.318 \text{ eV}$$

The diagonal core matrix elements over the atomic orbitals are

$$H_{11}^c = -I_1 - \sum_{\nu(\neq 1)} \hat{\gamma}_{1\nu} = -11.22 - 5.203 - 3.797 -$$

$$- 2.839 = -23.059 \text{ eV}$$

$$H_{22}^c = -I_2 - \sum_{\nu(\neq 2)} \hat{\gamma}_{2\nu} = -11.22 - 5.203 - 5.200 -$$

$$-3.797 = -25.423 \text{ eV}$$

From molecular symmetry follows

$$H_{11}^c = H_{44}^c \quad ; \qquad H_{22}^c = H_{33}^c$$

The molecular H_{11}^c integral then becomes

$$H_{11}^c = c_{11}\, c_{11}\, H_{11}^c + c_{11}\, c_{12}\, H_{12}^c + c_{12}\, c_{11}\, H_{21}^c +$$

$$+ c_{12}\, c_{12}\, H_{22}^c + c_{12}\, c_{13}\, H_{23}^c + c_{13}\, c_{12}\, H_{32}^c +$$

$$+ c_{13}\, c_{13}\, H_{33}^c + c_{13}\, c_{14}\, H_{34}^c + c_{14}\, c_{13}\, H_{43}^c +$$

$$+ c_{14}\, c_{14}\, H_{44}^c =$$

$$= -3.603 - 0.537 - 0.537 - 8.738 - 0.797 - 0.797 -$$

$$-8.738 - 0.537 - 0.537 - 3.603 = -28.426 \text{ eV}$$

62. $J_{mm} = \iint \varphi_m(1)\varphi_m(2) \dfrac{e^2}{r_{12}} \varphi_m(1)\varphi_m(2)\, d\tau_1\, d\tau_2 =$

$= \iint \left[\sum_\mu c_{m\mu} \chi_\mu(1)\right]^2 \dfrac{e^2}{r_{12}} \left[\sum_\nu c_{m\nu} \chi_\nu(2)\right]^2 d\tau_1\, d\tau_2 =$

$= \sum_\mu \sum_\nu c_{m\mu}^2\, c_{m\nu}^2 \underbrace{\iint \chi_\mu^2(1) \dfrac{e^2}{r_{12}} \chi_\nu^2(2)\, d\tau_1\, d\tau_2}_{\substack{\text{electronic repulsion}\\ \text{integral over AO's}}} =$

$= \sum_\mu \sum_\nu c_{m\mu}^2\, c_{m\nu}^2\, \gamma_{\mu\nu}$

If only repulsion integrals for one center and neighboring centers are retained, we obtain

$$J_{mm} \doteq 10.53 \sum_\mu c_{m\mu}^4 + 5.20 \sum_{\mu \neq \nu}\sum c_{m\mu}^2\, c_{m\nu}^2$$

In odd polyenes the singly occupied (m-th) level is a nonbonding MO, which means the expansion coefficients in the unstarred positions are zero:

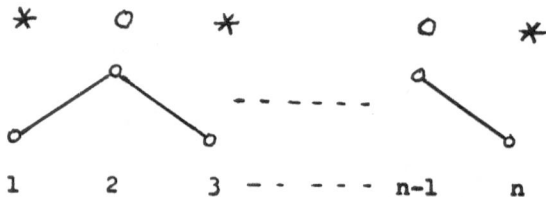

The coefficients for the starred positions can be determined from the normalization condition:

Number
of atoms

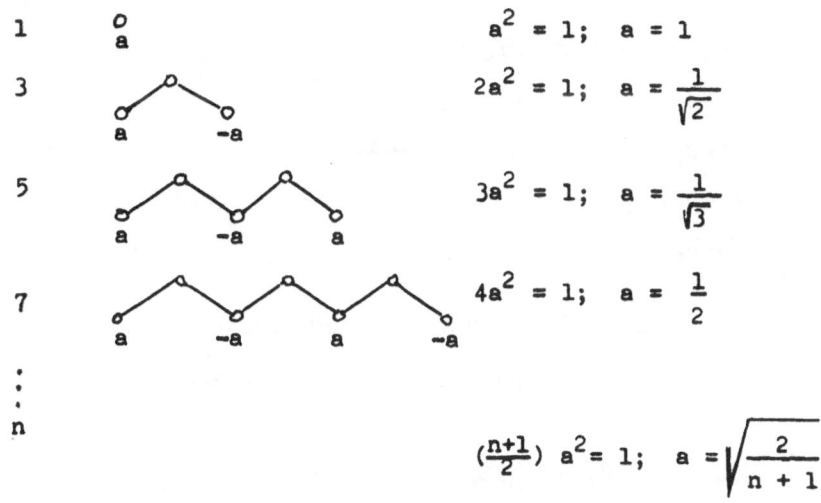

1	$\overset{\text{o}}{\underset{a}{}}$		$a^2 = 1; \quad a = 1$
3			$2a^2 = 1; \quad a = \dfrac{1}{\sqrt{2}}$
5			$3a^2 = 1; \quad a = \dfrac{1}{\sqrt{3}}$
7			$4a^2 = 1; \quad a = \dfrac{1}{2}$
$\vdots$			
n			$\left(\dfrac{n+1}{2}\right) a^2 = 1; \quad a = \sqrt{\dfrac{2}{n+1}}$

The J_{mm} integral becomes

Number of atoms	$c_{m\mu}$	$c^4_{m\mu}$	Number of starred atoms q	$J_{mm} = (qc^4_{m\mu})\ 10.53$
1	1	1	1	1×10.53
3	$1/\sqrt{2}$	1/4	2	$1/2 \times 10.53$
5	$1/\sqrt{3}$	1/9	3	$1/3 \times 10.53$
7	1/2	1/16	4	$1/4 \times 10.53$
$\vdots$				
n	$\sqrt{\dfrac{2}{n+1}}$	$4/(n+1)^2$	$\dfrac{n+1}{2}$	$\dfrac{2}{n+1} \times 10.53$

63. The Hückel-type matrix is obtained using the matrix elements

$$F_{\mu\mu} = -\tfrac{1}{2}(I_\mu + A_\mu)$$

$$F_{\mu\nu} = \tfrac{1}{2}(\beta^o_A + \beta^o_B)\, S_{\mu\nu}$$

Result:

	H(1s)	F(2s)	F(2p$_x$)	F(2p$_y$)	F(2p$_z$)
H(1s)	−7.176	−10.775	−8.014	0	0
F(2s)		−32.272	0	0	0
F(2p$_x$)			−11.080	0	0
F(2p$_y$)				−11.080	0
F(2p$_z$)					−11.080

64. The core matrix elements are defined as follows:

$$H_{\mu\mu} = -\tfrac{1}{2}(I_\mu + A_\mu) - (Z_A - \tfrac{1}{2})\,\gamma_{AA} - \sum_{B(\neq A)} Z_B\,\gamma_{AB}$$

$$H_{\mu\nu} = \tfrac{1}{2}(\beta_A^0 + \beta_B^0)S_{\mu\nu}$$

Z here is the core charge, in our case $Z_H = 1$ and $Z_F = 7$

$$H_{H(1s),H(1s)} = -7.176 - (1 - \tfrac{1}{2})\,20.407 - 7 \times 14.257 = -117.176$$

$$H_{F(2s),F(2s)} = -32.272 - (7 - \tfrac{1}{2})\,25.700 - 1 \times 14.257 =$$
$$= -213.582$$

$$H_{F(2p),F(2p)} = -11.080 - (7 - \tfrac{1}{2})\,25.700 - 1 \times 14.257 =$$
$$= -192.390$$

Core matrix elements $H_{\mu\nu}$

	H(1s)	F(2s)	F(2p$_x$)	F(2p$_y$)	F(2p$_z$)
H(1s)	−117.176	−10.775	−8.014	0	0
F(2s)		−213.582	0	0	0
F(2p$_x$)			−192.390	0	0
F(2p$_y$)				−192.390	0
F(2p$_z$)					−192.390

The $F_{\mu\nu}$ elements are obtained using the formulas

$$F_{\mu\mu} = H_{\mu\mu} + (P_{AA} - \tfrac{1}{2} P_{\mu\mu}) \gamma_{AA} + \sum_{B(\neq A)} P_{BB} \gamma_{AB}$$

$$F_{\mu\nu} = H_{\mu\nu} - \tfrac{1}{2} P_{\mu\nu} \gamma_{AB}$$

The electron density and bond order matrix is defined by

$$P_{\mu\nu} = \sum_{i=1}^{m} 2 \, c_{i\mu} \, c_{i\nu}$$

where the summation runs over the occupied MOs. The total electron densities on atoms, P_{AA}, are given by

$$P_{AA} = \sum_{\mu} P_{\mu\mu}$$

where μ runs over AOs belonging to atom A.

$$P_{\mu\nu}$$

	H(1s)	F(2s)	F($2p_x$)	F($2p_y$)	F($2p_z$)
H(1s)	0.6789	0.4226	0.8475	0	0
F(2s)		1.8648	−0.2711	0	0
F($2p_x$)			1.4563	0	0
F($2p_y$)				2.0000	0
F($2p_z$)					2.0000

$P_{HH} = 0.6789$

$P_{FF} = 1.8648 + 1.4563 + 2.0000 + 2.0000 = 7.3211$

$$F_{H(1s),H(1s)} = H_{H(1s),H(1s)} + (P_{HH} - \tfrac{1}{2} P_{H(1s),H(1s)}) \gamma_{HH} +$$
$$+ P_{FF} \gamma_{HF} =$$
$$= -117.176 + (0.6789 - 0.5 \times 0.6789) \, 20.407 +$$
$$+ 7.3211 \times 14.257 = -5.875$$

$$F_{F(2s),F(2s)} = H_{F(2s),F(2s)} + \left(P_{FF} - \frac{1}{2} P_{F(2s),F(2s)}\right) \Upsilon_{FF} +$$
$$+ P_{HH} \Upsilon_{HF} =$$
$$= -213.582 + (7.3211 - 0.5 \times 1.8648) \, 25.700 +$$
$$+ 0.6789 \times 14.257 = -39.711$$

$$F_{H(1s),F(2p_x)} = H_{H(1s),F(2p_x)} - \frac{1}{2} P_{H(1s),F(2p_x)} \Upsilon_{HF} =$$
$$= -8.014 - 0.5 \times 0.8475 \times 14.257 =$$
$$= -14.056$$

Similarly, the remaining matrix elements are obtained:

$$F_{\mu\nu}$$

	H(1s)	F(2s)	$F(2p_x)$	$F(2p_y)$	$F(2p_z)$
H(1s)	-5.875	-13.788	-14.056	0	0
F(2s)		-39.711	3.484	0	0
$F(2p_x)$			-13.269	0	0
$F(2p_y)$				-20.256	0
$F(2p_z)$					-20.256

65. LCI wave function is sought in the form

$$\Phi = c_0 \Psi_0 + c_1 \Psi_1 + c_2 \Psi_2 + c_3 \Psi_3 + c_4 \Psi_4$$

where $\Psi_0, \Psi_1, \Psi_2, \Psi_3, \Psi_4$ are Slater determinants representing the ground state and the four singly excited states: $1 \rightarrow 1'$, $1 \rightarrow 2'$, $2 \rightarrow 1'$, and $2 \rightarrow 2'$.

Using Equations (63) and (64) of the Appendix, the expression necessary for constructing the matrix elements are derived. The calculation of the F elements has already been shown in Exercise 59.

The required values are as follows:

$$F_{\mu\nu} \text{ (in eV)}$$

	1	2	3	4
1	-5.95500	-4.64485	0.00000	0.63483
2		-5.95500	-3.48138	0.00000
3			-5.95500	-4.64485
4				-5.95500

$$\sum_{\nu} F_{\mu\nu}\, c_{\ell\nu}$$

μ	$\ell = 1$	$\ell = 2$	$\ell = 1'$	$\ell = 2'$
1	-5.69050	-4.77165	-1.47336	0.34411
2	-3.71345	-7.40270	0.71409	-0.23884
3	3.71345	-7.40270	0.71409	0.23884
4	5.69050	-4.77165	-1.47336	-0.34411

$$F_{j\ell} = \sum_{\mu} c_{j\mu} \sum_{\nu} F_{\mu\nu}\, c_{\ell\nu}$$

	1	2		$1'$	$2'$
1	-9.60664	0	$1'$	-2.30338	0
2	0	-12.45316	$2'$	0	0.54316

	$1'$	$2'$
1	0	0.23638
2	-0.23638	0

$$(c_r c_{u'})_\mu$$

(r, u indicate indices of molecular orbitals, the primed symbols indicate antibonding MO)

μ	$11'$	$21'$	$12'$	$22'$	12	$1'2'$
1	0.36180	0.22361	0.22361	0.13820	0.22361	0.22361
2	-0.13820	-0.22361	-0.22361	-0.36180	0.22361	0.22361
3	0.13820	-0.22361	-0.22361	0.36180	-0.22361	-0.22361
4	-0.36180	0.22361	0.22361	-0.13820	-0.22361	-0.22361

$$\sum_{\nu} \widetilde{\gamma}_{\mu\nu} (c_s c_t')_{\nu}$$

μ	11'	22'	21'	12'
1	2.58824	0.55418	0.97702	0.97702
2	-0.22751	-1.73302	-1.50559	-1.50559
3	0.22751	1.73302	-1.50559	-1.50559
4	-2.58824	-0.55418	0.97702	0.97702

$$\sum_{\nu} \widetilde{\gamma}_{\mu\nu} (c_t \cdot c_u')_{\nu}$$

μ	1'1'	2'2'	1'2'
1	6.08072	5.10376	2.03416
2	5.43043	6.93594	1.50559
3	5.43043	6.93594	-1.50559
4	6.08072	5.10376	-2.03416

Note:

$$(rs|G|u't') = (rt'|G|u's) = \sum_{\mu} \sum_{\nu} c_{r\mu} c_{u'\mu} c_{t'\nu} c_{s\nu} \widetilde{\gamma}_{\mu\nu}$$

$$= \sum_{\mu} (c_r c_u')_{\mu} \sum_{\nu} \widetilde{\gamma}_{\mu\nu} (c_s c_t')_{\nu}$$

$(11|G|1'1') = (0.36180)(2.58824) + (-0.13820)(-0.22751) +$

$\qquad\qquad + (0.13820)(0.22751) + (-0.36180)(-2.58824) =$

$\qquad\qquad = 1.93574$

$(21|G|2'1') = 0.88000$

$(22|G|2'2') = 1.40720$

$(12|G|1'2') = 0.88000$

$(22|G|1'1') = 1.11026$

$(12|G|2'1') = 1.11026$

$(11|G|2'2') = 1.11026$

$(21|G|1'2') = 1.11026$

Note:

$$(ru'|G|st') = (rt'|G|su') = \sum_{\mu} \sum_{\nu} c_{r\mu} c_{s\mu} c_{t'\nu} c_{u'\nu} \tilde{\gamma}_{\mu\nu}$$

$$= \sum_{\mu} (c_r c_s)_{\mu} \sum_{\nu} (c_{t'} \cdot c_{u'})_{\nu} \tilde{\gamma}_{\mu\nu}$$

$(11'|G|11') = (0.36180)(6.08072) + (0.13820)(5.43043) +$

$\qquad\qquad + (0.13820)(5.43043) + (0.36180)(6.08072) =$

$\qquad = 5.90098$

$(21'|G|21') = 5.61018$

$(22'|G|22') = 6.42952$

$(12'|G|12') = 5.61018$

$(11'|G|22') = 1.58304$

Using Equations (60), (61), and (62) of the Appendix, one could compute the matrix elements and from them one could construct the LCI matrix. The matrix elements between states of different symmetry are zero. It is, therefore, possible to construct a separate matrix for the symmetric and for the antisymmetric states. The symmetry of the states can be determined from the symmetry of Hückel's molecular orbitals.

Ground state	$1 \rightarrow 1'$	$2 \rightarrow 1'$	$1 \rightarrow 2'$	$2 \rightarrow 2'$
A ——— 2'	———	———	——✳—	——✳—
S ——— 1'	—✳—	—✳—	———	———
A ✳✳ 1	✳—	✳✳	✳—	✳✳
S ✳✳ 2	✳✳	✳—	✳✳	✳—
S·S·A·A = S	S·S·A·S = A	S·A·A·S = S	S·S·A·A = S	S·A·A·A = A

Computation of the LCI matrix elements

$$< i \rightarrow j |H| k \rightarrow \ell >$$ symmetry states

Result

i = 1, j = 2′, k = 1, ℓ = 2′

$$<1 \rightarrow 2'|H|1 \rightarrow 2'> = F_{2'2'} - F_{11} - (12'|G|12') +$$
$$+ 2(11|G|2'2') =$$
$$= 0.54316 - (-9.60664) - 5.61018 +$$
$$+ 2(1.11026)$$ 6.76014

i = 1, j = 2′, k = 2, ℓ = 1′

$$<1 \rightarrow 2'|H|2 \rightarrow 1'> = -(22'|G|11') + 2(22'|G|1'1)$$
$$= -1.58304 + 2(1.11026)$$ 0.63748

i = 1, j = 2′, k = ℓ

$$<1 \rightarrow 2'|H|V_0> = \sqrt{2}\, F_{12'} = \sqrt{2}\,(0.23638)$$ 0.33429

(V_0 indicates a single determinant function of the ground state.)

i = 2, j = 1′, k = 2, ℓ = 1′

$$<2 \rightarrow 1'|H|2 \rightarrow 1'> = F_{1'1'} - F_{22} - (21'|G|21') +$$
$$+ 2(21'|G|1'2) =$$
$$= -2.30338 - (-12.45316) -$$
$$- 5.61018 + 2(1.11026)$$ 6.76012

i = 2, j = 1′, k = 1, ℓ = 2′

$$<2 \rightarrow 1'|H|1 \rightarrow 2'> = -(11'|G|22') + 2(11'|G|2'2) =$$
$$= -1.58304 + 2(1.11026$$ 0.63748

Result

$\underline{i = 2, \ j = 1', \ k = \ell}$

$<2 \rightarrow 1' | H | V_o> = \sqrt{2} \ F_{21'} = \sqrt{2} \ (-0.23638)$ -0.33429

$<i \rightarrow j \ H \ k \rightarrow \ell>$ antisymmetry states

$\underline{i = 1, \ j = 1', \ k = 1, \ \ell = 1'}$

$<1 \rightarrow 1' | H | 1 \rightarrow 1'> = F_{1'1'} - F_{11} - (11'|G|11') +$

$+ 2(11' \ G \ 1'1) =$

$= -2.30388 - (-9.60664) - 5.90098 +$

$+ 2(1.93574)$ 5.27376

$\underline{i = 1, \ j = 1', \ k = 2, \ \ell = 2'}$

$<1 \rightarrow 1' | H | 2 \rightarrow 2'> = - (21'|G|12') + 2(21'|G|2'1) =$

$= -1.58304 + 2(0.88000)$ 0.17696

$\underline{i = 2, \ j = 2', \ k = 2, \ \ell = 2'}$

$<2 \rightarrow 2' | H | 2 \rightarrow 2'> = F_{2'2'} - F_{22} - (22'|G|22') +$

$+ 2(22'|G|2'2) =$

$= 0.54316 - (-12.45316) - 6.42952 +$

$+ 2(1.40720)$ 9.38120

$\underline{i = 2, \ j = 2', \ k = 1, \ \ell = 1'}$

$<2 \rightarrow 2' | H | 1 \rightarrow 1'> = - (12'|G|21') + 2(12'|G|1'2) =$

$= -1.58304 + 2(0.88000)$ 0.17696

$< i \rightarrow j|H|k \rightarrow \ell >$ LCI matrix (symmetric states)

	$1 \rightarrow 2'$	$2 \rightarrow 1'$	V_o
$1 \rightarrow 2'$	6.76014	0.63748	0.33429
$2 \rightarrow 1'$	0.63748	6.76012	-0.33429
V_o	0.33429	-0.33429	0

$< i \rightarrow j|H|k \rightarrow \ell >$ LCI matrix (antisymmetric states)

	$1 \rightarrow 1'$	$2 \rightarrow 2'$
$1 \rightarrow 1'$	5.27376	0.17696
$2 \rightarrow 2'$	0.17696	9.38120

Computation Results

Energies of singlet states E and the expansion coefficients of normalized LCI wave functions:

E(eV)	00	$11'$	$12'$	$21'$	$22'$	
-0.03629	-0.99707	–	0.05412			ground state
5.26610	–	0.99908	–	–	-0.04295	1.excited state
6.15893	0.07653	–	0.70503	-0.70503	–	2.excited state
7.39763	0.00000	–	0.70711	0.70711	–	3.excited state
9.38875	–	0.04295	–	–	0.99908	4.excited state

66. Orbital energy in β units $(E = \alpha + k\beta)$ can be transformed into orbital energy in γ units using the expression

$$E_i = \alpha + \frac{k_i}{1 + k_i \times 0.25}\, \gamma$$

No.	Hydrocarbon	S = 0 (β-units) E(N→V₁)	S = 0.25 (γ-units)			Experimental data (band)							
			HOMO	LFMO	E(N→V₁)	α mμ	α kcm⁻¹	p mμ	p kcm⁻¹	β mμ	β kcm⁻¹	β' mμ	β' kcm⁻¹
I	Benzene	2.000	0.800	-1.333	2.133	260	38.5	210	47.6	-	-	-	-
II	Naphthalene	1.236	0.535	-0.731	1.266	311	32.2	285	35.1	221	45.3	-	-
III	Anthracene	0.828	0.375	-0.462	0.837	-	-	375	26.7	250	40.0	-	-
IV	Tetracene	0.590	0.275	-0.369	0.644	-	-	471	21.2	274	36.5	-	-
V	Phenanthrene	1.210	0.525	-0.713	1.238	345	29.0	292	34.2	250	40.0	210	47.6
VI	Chrysene	1.040	0.460	-0.598	1.058	360	27.8	319	31.4	267	37.4	220	45.5
VII	Pyrene	0.890	0.401	-0.501	0.902	372	26.9	334	29.9	272	36.8	241	41.5
VIII	Benzo[c]-phenanthrene	1.135	0.497	-0.662	1.159	372	26.9	315	31.8	281	35.6	218	45.9
IX	Benz[e]-anthracene	0.905	0.406	-0.510	0.916	385	26.0	359	27.8	290	34.5	227	44.0

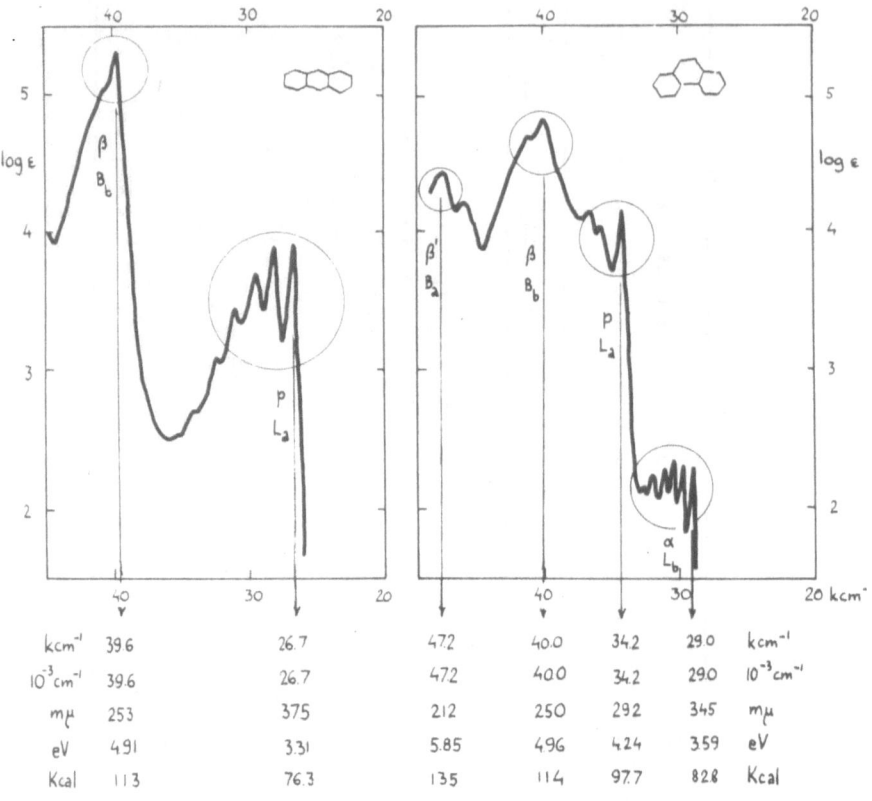

Fig. 8. Absorption curves of anthracene and phenanthrene.

For example, in the case of naphthalene the energies of HOMO and LFMO have the following values:

$$E_1 = \alpha + 0.618\,\beta$$

$$E_{-1} = \alpha - 0.618\,\beta$$

By substituting into the above formula, we obtain

$$E_1 = \alpha + \frac{0.618}{1 + 0.618 \times 0.25}\,\gamma = \alpha + 0.535\,\gamma$$

$$E_{-1} = \alpha + \frac{(-0.618)}{1 - 0.618 \times 0.25}\,\gamma = \alpha - 0.731\,\gamma$$

(Note that by passing from β to γ units the diagram of the orbital

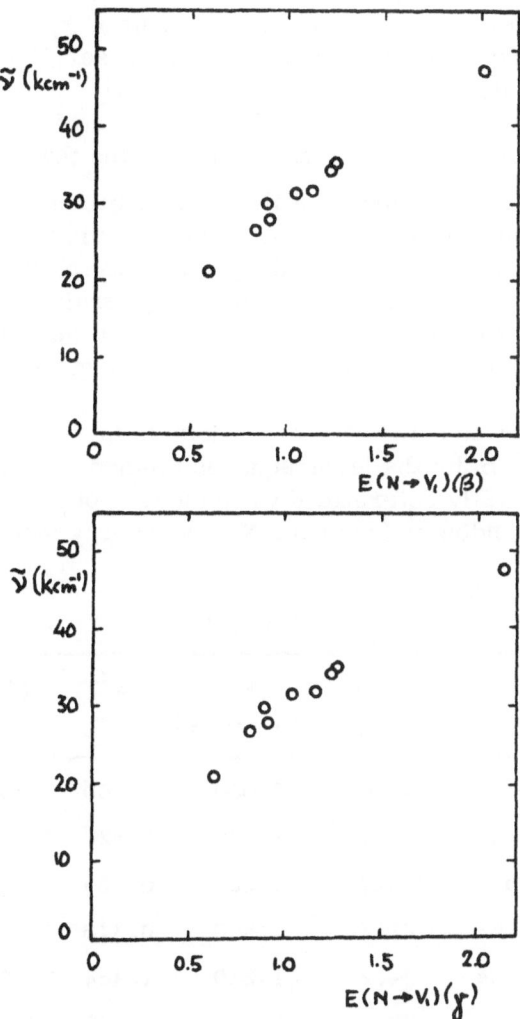

Fig. 9. Wave numbers of the first intense bands
(p bands) plotted against the $N \rightarrow V_1$ transition energies in β and γ units.

levels of an alternant hydrocarbon loses its symmetry.) By adding the orbital energies computed in this manner, we obtain the
energies of the $N \rightarrow V_1$ transitions in γ units (see table p. 149).
The procedure of the empirical assignment of α, p, β, and β' bands
can be seen from Fig. 8, which shows the absorption curves of
phenanthrene (typical example of an absorption curve with α, p, β,
and β' bands) and anthracene (typical example of an absorption curve
in which the p band lies at longer wavelengths than the α band).

Treat the absorption curves of the other hydrocarbons in a similar
way. The results of this treatment are given in the table and in Fig.
9, where the wave numbers of the p bands are plotted against the
energies $E(N \rightarrow V_1)$ in β and γ units. The straight line fitted to the
points does not pass through the origin as a result of oversimplifi-
cations in the HMO method (neglect of the electron repulsion).

67. The problem is to determine the constants (a, b) of the regression
line (y = ax + b) by the least-squares method. In this case y denotes
the experimentally determined wave number of the maximum and
x the corresponding value of the $N \rightarrow V_1$ energy (see Table I).

Table I

Substance	y $\tilde{\nu}(kcm^{-1})$	x $E(N \rightarrow V_1)$	x^2	y^2	xy
Benzene	47.6	2.000	4.000	2265.76	95.2
Naphthalene	35.1	1.236	1.527	1232.10	43.38
Anthracene	26.7	0.828	0.685	712.89	22.11
Tetracene	21.2	0.590	0.348	449.44	12.51
Phenanthrene	34.2	1.210	1.464	1169.64	41.38
Chrysene	29.9	0.890	0.792	894.01	26.61
Benzo[c]-phenanthrene	31.8	1.135	1.288	1011.24	36.09
Benz[a]-anthracene	27.8	0.905	0.819	772.84	25.16

$$\Sigma x = 9.834 \; ; \; \Sigma y = 285.7 \; ; \; \Sigma xy = 335.09$$

$$\Sigma x^2 = 12.00 \; ; \; \Sigma y^2 = 9493.88 \; ; \; (\Sigma x)^2 = 96.707$$

$$(\Sigma y)^2 = 81624.49$$

$$\bar{x} = \frac{\Sigma x}{N} = \frac{9.834}{9} = 1.093$$

$$\bar{y} = \frac{\Sigma y}{N} = \frac{285.7}{9} = 31.744$$

$$\Sigma (x - \bar{x})^2 = \Sigma x^2 - \frac{(\Sigma x)^2}{N} = 12.00 - 10.745 = 1.255$$

$$\Sigma (y - \bar{y})^2 = \Sigma y^2 - \frac{(\Sigma y)^2}{N} = 9493.88 - 9069.38 = 424.5$$

The equation of the regression line is

$$y = ax + b$$

where

$$a = \frac{\Sigma xy - \dfrac{\Sigma x \Sigma y}{N}}{\Sigma x^2 - \dfrac{(\Sigma x)^2}{N}}$$

$$a = \frac{335.09 - \dfrac{(9.834)(285.7)}{9}}{12.00 - \dfrac{96.707}{9}} = 18.26$$

$$b = \bar{y} - a\bar{x}$$

$$b = 31.774 - 18.26 \times 1.093 = 11.82$$

$$y = 18.26x + 11.82$$

In this case we may write

$$\tilde{y} = 18.26 \ E(N \longrightarrow V_1) + 11.82$$

For the plot, see Fig. 10.

The correlation coefficient is

$$r = \frac{\sum xy - \dfrac{\sum x \sum y}{N}}{\sqrt{\sum (x - \bar{x})^2 \sum (y - \bar{y})^2}}$$

$$r = \frac{335.09 - \dfrac{(9.834)(285.7)}{9}}{\sqrt{1.255 \times 424.5}} = 0.993$$

From this value of the correlation coefficient, with a view to the number of points (N = 9), and from the data in Table II it is clear that the correlation is statistically significant when testing at a 1% significance level.

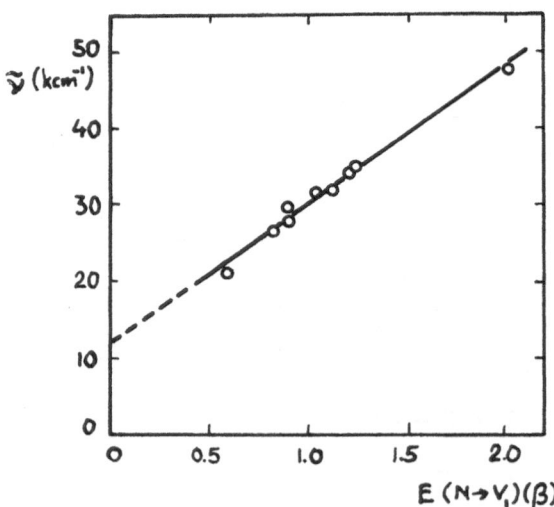

Fig. 10. Least-squares plot of the experimental wave numbers against the $N \rightarrow V_1$ transition energies.

Table II
Critical Value of the Correlation Coefficient

(A correlation is statistically significant if for the given value of n the calculated value of r^* is equal to or greater than the tabulated critical value)

n	r_5	r_1	n	r_5	r_1
3	0.997	1.000	18	0.468	0.590
4	0.950	0.990	19	0.456	0.575
5	0.878	0.959	20	0.444	0.561
6	0.811	0.917	21	0.433	0.549
7	0.755	0.875	22	0.423	0.537
8	0.707	0.834	25	0.395	0.511
9	0.666	0.798	30	0.361	0.463
10	0.632	0.765	35	0.334	0.429
11	0.602	0.735	40	0.312	0.401
12	0.576	0.708	45	0.294	0.379
13	0.553	0.684	50	0.279	0.361
14	0.532	0.661	60	0.254	0.330
15	0.514	0.641	70	0.235	0.306
16	0.497	0.623	80	0.220	0.286
17	0.482	0.616	90	0.207	0.270
			100	0.197	0.256

$^*r_1$ and r_5 reflect significance at probability levels of 1 and 5%, respectively.

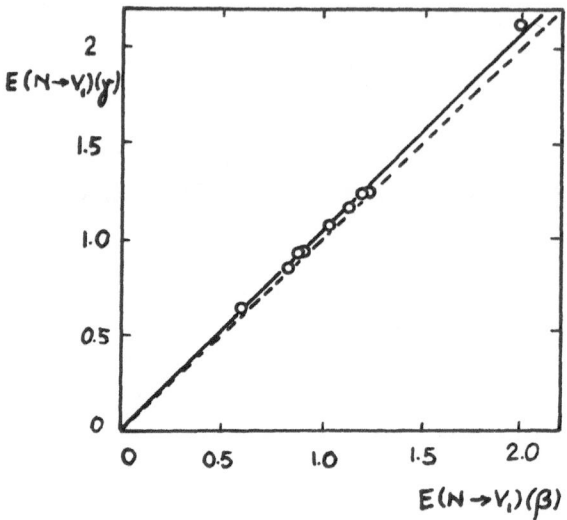

Fig. 11. $N \to V_1$ transition energies in γ units plotted
against the same quantities in β units.

In **Fig. 11** the $N \to V_1$ energies in β units are plotted against the
same characteristics in γ units. The figure indicates that the
points deviate only little from a straight line passing through the
origin and having nearly unit slope. This means that not only the
order of the $N \to V_1$ energies in β and γ units is parallel, but also
that their absolute values are very similar.

68. For the change in the energy of the $N \to V_1$ transition due to a change
in the Coulomb integral in the position μ of $\delta \alpha_\mu$ the following equation, based on the first-order pertubation treatment, is valid (the
$N \to V_1$ transition corresponds to $i \to j$ excitation):

$$\delta E(N \to V_1) = (c_{j\mu}^2 - c_{i\mu}^2)\delta \alpha_\mu ,$$

where $c_{i\mu}$ stands for the expansion coefficient in the i-th molecular
orbital belonging to the atomic orbital μ. In our particular case,

$$\delta E(N \to V_1) = (0.546 - 0) \; 13.8 \; \text{kK} = 7.53 \; \text{kK} .$$

The expected shift of the first absorption band when passing from
I to II therefore amounts to 7.53 kK.

If we correct the theoretical excitation energy for I (17.3 kK) by this value, we get the expected wave number of the first absorption band in II:

$$17.3 \text{ kK} + 7.5 \text{ kK} = 24.8 \text{ kK} \qquad (403 \text{ m}\mu)$$

This value corresponds to the yellow color.

The pronounced bathochromic shift when passing from II to III may be explained by the departure of the dimethylamino group from coplanarity owing to the steric hindrance. (Consider that the $\delta\alpha_\mu$ constants in fact also cover mesomeric effects.) Thus, one would expect an only slightly perturbed parent system I, which, according to the theoretical prediction, should be blue.

69. Maxima and minima on the fluorescence polarization curve indicate the positions of the first four absorption bands. They clearly show that the band in the region of 30,000 cm^{-1} corresponds to two electronic transitions (Fig. 12). Moreover those data say that the first and third transitions are parallelly polarized; the same is true about the second and fourth transitions.

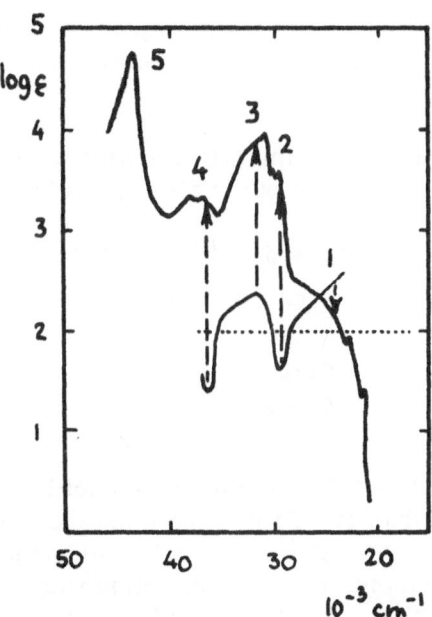

Fig. 12. Use of the fluorescence polarization curve of acenaphthylene for assignment of electronic bands.

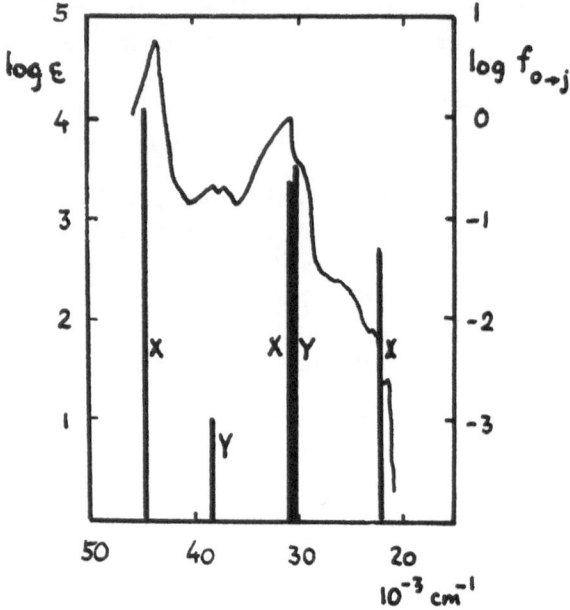

Fig. 13. Electronic absorption curve of acenaphthylene
and results of the LCI-SCF calculations.

In the following table the primarily calculated values are summarized together with some other simply derived data:

E_i	$\nu(10^{-3}\ cm^{-1})$	$f_{o \rightarrow j}$	$\log f_{o \rightarrow j}$	$\cos \theta$	$\theta(degree)$
2.75	22.2	0.04	−1.4	1	0
3.75	30.3	0.31	−0.51	0	90
3.83	30.9	0.22	−0.66	1	0
4.75	38.4	0.001	−3	0	90
5.56	44.9	1.21	0.08	1	0

There is apparently a fair overall agreement between theoretical
and experimental data (Fig. 13). The existence of two transitions
in the region of 30,000 cm^{-1} is strongly supported by CI data. Somewhat poorer agreement as far as the intensities are concerned is
easily understandable: the theoretical $f_{o \rightarrow j}$ values should be com-
pared with experimental oscillator strengths and not with molar
extinction coefficients. However, these data are obtainable only
rarely because of the strong overlap of bands in electronic spectra.

It appears that the first, second, and fourth bands are due to nearly pure transitions (i.e., that in the respective excited-state wave
function a certain configuration predominates). Obviously, the
first band can be ascribed to the $N \rightarrow V_1$ transition.

70.

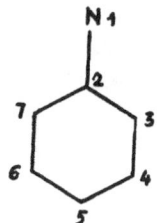

$$\tilde{\gamma}_{\mu\nu} \text{ (eV)}$$

	1	2	3	4	5	6	7
1	18.000	5.997	4.203	3.060	2.768	3.060	4.203
2		10.530	5.203	3.797	3.455	3.797	5.203
3			10.530	5.203	3.797	3.455	3.797
4				10.530	5.203	3.797	3.455
5					10.530	5.203	3.797
6						10.530	5.203
7							10.530

$$(c_r c_u)_\mu$$

μ	22	11	1'1'	2'2'	11'
1	0	0.43818	0	0.05413	0
2	0	0.09897	0	0.32625	0
3	0.25290	0.12186	0.24710	0.06077	-0.17352
4	0.24710	0.01805	0.25290	0.08926	-0.06756
5	0	0.18305	0	0.31958	0
6	0.24710	0.01805	0.25290	0.08926	0.06756
7	0.25290	0.12186	0.24710	0.06077	0.17352

μ	21′	22′	12′	12	1′2′
1	0	0	0.15400	0	0
2	0	0	0.17969	0	0
3	0.25000	0.12397	-0.08606	-0.17555	0.12254
4	-0.25000	0.14851	0.04014	0.06678	-0.15024
5	0	0	-0.24188	0	0
6	-0.25000	-0.14851	0.04014	-0.06678	0.15024
7	0.25000	-0.12397	-0.08606	0.17555	-0.12254

$$\sum_\nu \mathcal{T}_{\mu\nu} \, (c_s c_t)_\nu$$

μ	11′	21′	12′	22′	1′1′	2′2′	1′2′
1	0	0.57150	2.70232	0	3.62486	4.87256	0
2	0	0.70300	1.38926	0	4.49184	6.17439	0
3	-1.28640	1.41725	-0.21513	1.09429	5.72981	4.78190	0.54684
4	-0.75819	-1.41725	-0.27501	1.21662	5.72981	4.87216	-0.79737
5	0	-0.70300	-1.73574	0	4.49184	6.03253	0
6	0.75819	-1.41725	-0.27501	-1.21662	5.72981	4.87216	0.79737
7	1.28640	1.41725	-0.21513	-1.09429	5.72981	4.78190	-0.54684

$$(rt|G|su) = \sum_\mu (c_r c_s)_\mu \sum_\nu (c_t c_u)_\nu \, \mathcal{T}_{\mu\nu}$$

$$(12'|G|12') = 5.191$$

$$(12'|G|2'1) = 1.101$$

$$(11'|G|22') = -0.298$$

$$(11'|G|2'2) = 0.029$$

$$(21'|G|21') = 5.730$$

$$(21'|G|1'2) = 1.417$$

$$(11'|G|11') = 4.458$$

$$(11'|G|1'1) = 0.549$$

$$(21'|G|2'1) = -0.544$$

$$(22'|G|22') = 4.826$$

$$(22'|G|2'2) = 0.633$$

Now all expressions needed for setting up the LCI matrix are at our disposal. Since the calculation is based on the SCF molecular orbitals, the formulas (61) and (62) of the Appendix become simpler owing to the relations

$$F_{ij} = 0 \qquad \text{(for } i \neq j)$$

$$F_{ii} = E_i \qquad \text{(SCF orbital energies)}$$

We employ symmetry in the same way as in Exercise 65. Thus, the symmetric and antisymmetric states can be treated separately.

$$< i \to j |H| k \to \ell > \text{ , symmetric states}$$

<u>$i = 1, \ j = 2'; \ k = 1, \ \ell = 2'$</u>   Result

$$< 1 \to 2'|H|1 \to 2' > = F_{2'2'} - F_{11} - (12'|G|12') +$$

$$+ \ 2(12'|G|2'1) =$$

$$= -0.952 - (-9.184) - 5.191 +$$

$$+ \ 2(1.101) \qquad\qquad\qquad 5.243$$

<u>$i = 1, \ j = 2'; \ k = 2, \ \ell = 1'$</u>

$$< 1 \to 2'|H|2 \to 1' > = -(22'|G|11') + 2(22'|G|11'1) =$$

$$= -0.298 + 2(0.029) \qquad\qquad 0.356$$

<u>$i = 2, \ j = 1'; \ k = 2, \ \ell = 1'$</u>

$$< 2 \to 1'|H|2 \to 1' > = F_{1'1'} - F_{22} - (21'|G|21') +$$

$$+ \ 2(21'|G|1'2) =$$

$$= -1.152 - (-10.423) - 5.730 +$$

$$+ \ 2(1.417) \qquad\qquad\qquad 6.375$$

<u>$i = 1, \ j = 2', \ k = \ell$</u>

$$< 1 \to 2'|H|V_o > = \sqrt{2} \ F_{12'} \qquad\qquad\qquad 0$$

Result

$\underline{i = 2, \ j = 1', \ k = \ell}$

$< 2 \rightarrow 1' |H| V_o > \ = \sqrt{2} \ F_{21'}$ 0

LCI matrix, symmetric states

	$1 \rightarrow 2'$	$2 \rightarrow 1'$	V_o
$1 \rightarrow 2'$	5.243	0.356	0
$2 \rightarrow 1'$	0.356	6.375	0
V_o	0	0	0

Thus, the secular polynomial is

$$(5.243 - E)(6.375 - E) - (0.356)^2 = 0$$

and yields energy values of 5.140 and 6.477 eV.

$$< i \rightarrow j |H| k \rightarrow \ell > \ , \ \text{antisymmetric states}$$

$\underline{i = 1, \ j = 1'; \ k = 1, \ \ell = 1'}$

$< 1 \rightarrow 1' |H| 1 \rightarrow 1' > \ = F_{1'1'} - F_{11} - (11'|G|11') +$

$\qquad\qquad\qquad\qquad + 2(11'|G|1'1) =$

$\qquad\qquad\qquad = -1.152 - (-9.184) - 4.458 +$

$\qquad\qquad\qquad\qquad + 2(0.549)$ 4.668

$\underline{i = 1, \ j = 1'; \ k = 2, \ \ell = 2'}$

$< 1 \rightarrow 1' |H| 2 \rightarrow 2' > \ = \ - (21'|G|12') + 2(21'|G|2'1)$

$\qquad\qquad\qquad = 0.298 + 2(-0.544)$ −0.790

$\underline{i = 2, \; j = 2'; \; k = 2, \; \ell = 2'}$ Result

$$\langle 2 \rightarrow 2' | H | 2 \rightarrow 2' \rangle = F_{2'2'} - F_{22} - (22'|G|22') +$$

$$+ \; 2(22'|G|2'2) =$$

$$= \; -0.952 - (-10.423) - 4.826 +$$

$$+ \; 2(0.633) \qquad\qquad\qquad 5.911$$

LCI matrix, antisymmetric states

	$1 \rightarrow 1'$	$2 \rightarrow 2'$
$1 \rightarrow 1'$	4.668	-0.790
$2 \rightarrow 2'$	-0.790	5.911

The secular polynomial for the antisymmetric states is

$$(4.668 - E)(5.911 - E) - (-0.790)^2 = 0$$

and the corresponding energy values are 4.282 and 6.295 eV.

Computation Results

Energies of the singlet states and expansion coefficients of the
normalized LCI wave functions:

E (eV)	E (kK)		11'	12'	21'	22'	
0	0	1.000	-	-	-	-	ground state
4.284	34.56	-	0.899	-	-	0.436	1st excited state
5.140	41.47	-	-	0.959	-0.281	-	2nd excited state
6.295	50.79	-	-0.436	-	-	0.899	3rd excited state
6.477	52.26	-	-	0.281	0.959	-	4th excited state

Calculated transition energies indicate the positions of the absorption maxima in electronic spectra; however for the interpretation
of the spectral intensities one needs to calculate also the oscillator
strengths. These values are presented in the following table.

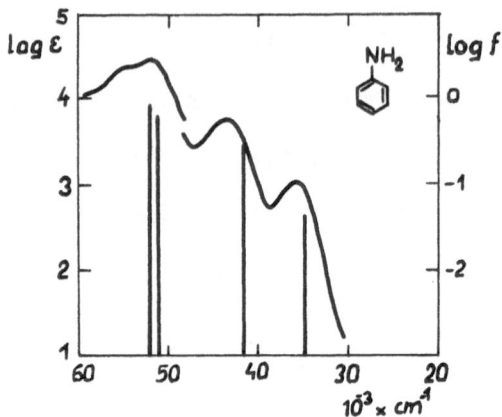

Fig. 14. Electronic absorption curve of aniline
and results of the LCI-SCF calculations.

Logarithms of the Oscillator Strengths

Transition	log f
1	−1.371
2	−0.470
3	−0.100
4	−0.029

The results of the LCI-SCF calculation are illustrated in Fig. 14.

71. For the FEMO orbital energies of linear polyenes the following
expression holds (see Appendix):

$$E_n = \frac{n^2 h^2}{8\,m l^2}$$

It is assumed that in the ground state N π-electrons of the even
linear polyene occupy N/2 of the lowest orbitals. The first intense
band, therefore, is connected with the jump of an electron from
the N/2 level to the (N/2 + 1) level. If we identify the length of

tion energy is obtained:

$$\Delta E = \frac{(N/2 + 1)^2 \, h^2}{8m \left[1.4(N - 1)\right]^2} - \frac{(N/2)^2 \, h^2}{8m \left[1.4(N - 1)\right]^2}$$

$$\Delta E = \frac{h^2 \, (N + 1)}{8m \, (1.4)^2 \, (N - 1)^2}$$

Since $\Delta E = h\nu$ and $\lambda = c/\nu$, the corresponding wavelength (in Å) is

$$\lambda = \frac{8m \, (1.4)^2 \, c(N - 1)^2}{h \, (N + 1)}$$

and if all constants are combined into a single one, A

$$\lambda = A \, \frac{N^2 - 2N + 1}{N + 1}$$

The HMO excitation energies of the $N \rightarrow V_1$ transition in β-units are:

$$E = \left[2 \cos \frac{\pi(N/2 + 1)}{N + 1} - 2 \cos \frac{\pi N/2}{N + 1} \right] \beta$$

We shall use the relation

$$\cos x - \cos y = -2 \sin \frac{x+y}{2} \sin \frac{x-y}{2}$$

to find

$$\Delta E = -4 \sin \frac{\pi}{2N + 2} \, \beta$$

If N is large,

$$\sin \frac{\pi}{2N + 2} \doteq \frac{\pi}{2N + 2}$$

so that

$$\Delta E = -\frac{2\pi}{N + 1} \, \beta$$

We again include all the constants in one and obtain the following expression for the wavelength:

$$\lambda = B \ (N + 1)$$

For long linear polyenes, according to both methods, the wavelength of the absorption maximum of the first intense band in the electronic spectrum is proportional to the number of p_z atomic orbitals.

72. The transition moment in the HMO method is defined by

$$\vec{\lambda} = \sqrt{2} \sum_{\mu} c_{\mu}^{1} \ c_{\mu}^{-1} \ \vec{r}_{\mu}$$

where c_{μ}^{1} (c_{μ}^{-1}) denotes the expansion coefficients with the μ-th atomic orbital in the highest occupied (the lowest free) molecular orbital.

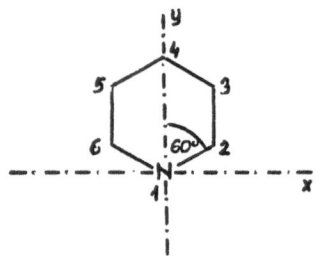

Transition $1 \longrightarrow 1'$

μ	x	y	c_{μ}^{1}	c_{μ}^{-1}	$c_{\mu}^{1} c_{\mu}^{-1} x$	$c_{\mu}^{1} c_{\mu}^{-1} y$
1	0	0	0.000	−0.546	0.000	0.000
2	1.21	0.7	−0.500	0.366	−0.221	−0.128
3	1.21	2.1	−0.500	0.238	−0.144	−0.250
4	0	2.8	0.000	−0.566	0.000	0.000
5	−1.21	2.1	0.500	0.238	−0.144	0.250
6	−1.21	0.7	0.500	0.366	−0.221	0.128

$$|\vec{\lambda}| = \sqrt{\lambda_x^2 + \lambda_y^2}$$

$$\lambda_x = 2(-0.221 - 0.144 - 0.144 - 0.221) = -\sqrt{2} \times 0.73$$

$$\lambda_y = 2(-0.128 - 0.250 + 0.250 + 0.128) = 0$$

$$|\vec{\lambda}| = |\lambda_x|$$

Transition $1 \rightarrow 2'$

μ	x	y	c_μ^1	c_μ^{-2}	$c_\mu^1 c_\mu^{-2} x$	$c_\mu^1 c_\mu^{-2} y$
1	0	0	0.000	0.000	0.000	0.000
2	1.21	0.7	-0.500	0.500	-0.302	-0.175
3	1.21	2.1	-0.500	-0.500	0.302	0.525
4	0	2.8	0.000	0.000	0.000	0.000
5	-1.21	2.1	0.500	0.500	-0.302	0.525
6	-1.21	0.7	0.500	-0.500	0.302	-0.175

$$\lambda_x = \sqrt{2}\,(-0.302 + 0.302 - 0.302 + 0.302) = 0$$

$$\lambda_y = \sqrt{2}\,(-0.175 + 0.525 + 0.525 - 0.175) = \sqrt{2} \times 0.70$$

$$|\vec{\lambda}| = |\lambda_y|$$

The vector $\vec{\lambda}$ denotes the polarization of the absorbed radiation. For $\pi \rightarrow \pi^*$ transitions it lies in the plane of the molecule. In our case the polarization directions are:

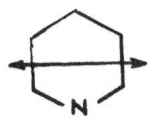

Transition $1 \rightarrow 1'$

Transition $1 \rightarrow 2'$

73.

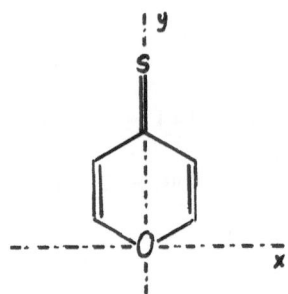

μ	x	y	$c_\mu^1 \, c_\mu^{-1}$ x	$c_\mu^1 \, c_\mu^{-1}$ y
1	0	0	0.0000	0.0000
2	1.21	0.70	0.0998	0.0573
3	1.21	2.10	0.0174	0.0300
4	0	2.80	0.0000	-0.0087
5	-1.21	2.10	-0.0174	0.0300
6	-1.21	0.70	-0.0998	0.0573
7	0	4.20	0.0000	-1.1903

$$\left| \vec{\lambda} \right| = \sqrt{\lambda_x^2 + \lambda_y^2}$$

$$\lambda_x = \sqrt{2} \, (0.0998 + 0.0174 - 0.0174 - 0.0998) = 0$$

$$\lambda_y = \sqrt{2} \, (0.0573 + 0.0300 - 0.0087 + 0.0300 - 0.0573 -$$

$$- 1.1903) =$$

$$= - \sqrt{2} \times 1.02$$

$$\left| \vec{\lambda} \right| = \left| \lambda_y \right|$$

Direction of polarization:

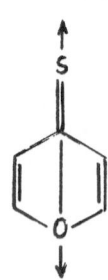

74. The theoretical excitation energy can be expressed as the sum of HMO excitation energy and the contribution originating from the electronic repulsion. The latter is a quantity dependent on the structure in the sense that it differs for various structural types, e.g., for cyclic hydrocarbons and linear hydrocarbons or for neutral hydrocarbons and hydrocarbon ions.[24] Since within one class of substances of a certain type this contribution changes parallel to HMO excitation energies, linear relationships are also obtained for HMO data.

75. The spin density for a given position can be represented by the square of the expansion coefficient (for this position) of the molecular orbital which contains the odd unpaired electron (in this case LFMO). Therefore, the hyperfine splitting constants are correlated with the squares of the expansion coefficients. The appropriate coefficients are given on pp. 193-195.

Substance	Position (μ)	c_μ^2
Naphthalene	1	0.181
	2	0.070
Anthracene	1	0.097
	2	0.048
	3	0.193
Tetracene	1	0.056
	2	0.034
	5	0.147
Perylene	1	0.083
	2	0.013
	3	0.108
Pyrene	1	0.136
	2	0.000
	4	0.087

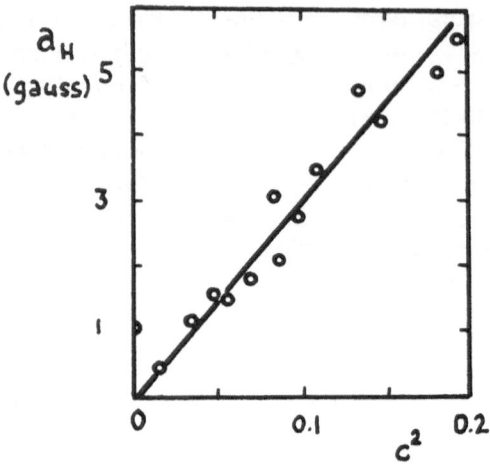

Fig. 15. Hyperfine splitting constants (a_H)
plotted against the squares of the respective
HMO expansion coefficients.

The plot of Fig. 15 indicates the following empirical equation for
this correlation:

$$a_{H(gauss)} = 30 \times c_{1\mu}^2$$

N o t e : If the value of the expansion coefficient is zero or near to
zero, it is not possible to consider the theoretical value to be a
reliable measure of the spin density (the case of position 2 with
pyrene); in the HMO method it is impossible to express the negative spin density. If it is desired to remove this deficiency, the
LCI or unrestricted SCF method must be used.[25]

76. (a) Naphthalene has four equivalent protons in α positions and four
equivalent protons in β positions. By hyperfine splitting due to
the α-protons 5 lines are created. Each of these lines, however,
is split by β protons into 5 lines. The ESR spectrum of the naphthalene radical anion, therefore, consists of 25 lines.

(b)

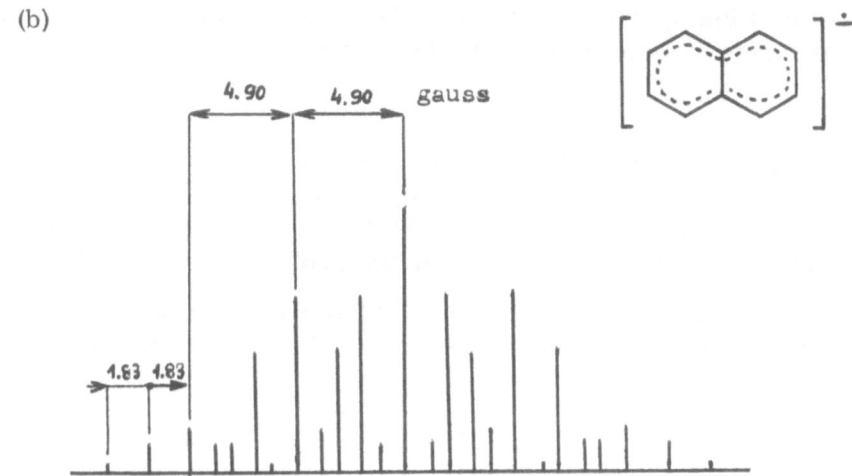

4.90   4.90   gauss

1.83 1.83

(c) 26.92 gauss.

(d) The hyperfine splitting constants are proportional to the spin densities, which can be approximated by the squares of HMO expansion coefficients of that molecular orbital which is occupied by the unpaired electron. For alternant hydrocarbons the squares of the HMO expansion coefficients of HOMO and LFMO are the same (pairing properties).

77.

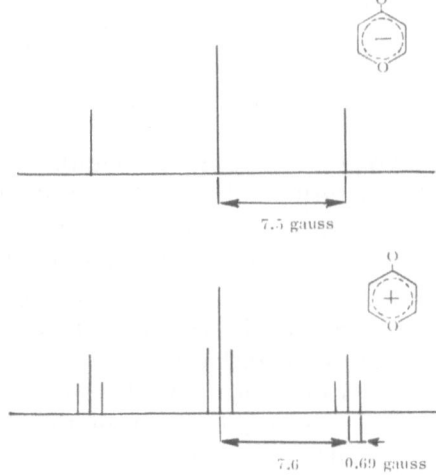

7.5 gauss

7.6   0.69 gauss

78. The following relation is used to interpret ESR experimental data in terms of molecular orbital methods:

$$a = Q\rho$$

Here a is the hyperfine splitting constant; Q is the proportionality constant, and ρ is the spin density. In the present case the constant a_1 characterizes splitting by four equivalent protons in the positions 1 and 3, a_2 splitting by one proton in the position 2.

(a) In the HMO method the spin densities are represented by the squares of the expansion coefficients of the singly occupied molecular orbital:

$$a_\mu = Q\ c_\mu^2$$

In case of the allyl radical the molecular orbital is nonbonding (orbital energy $E = \alpha + k\beta; k = 0$)

The squares of the expansion coefficients of this nonbonding molecular orbital are as follows:

μ	1	2	3
$c_{1\mu}^2$	0.500	0.000	0.500

The hyperfine splitting constant a_2 should, therefore, be zero. This is a typical case of failure of the HMO method. Such failures occur frequently if the value of the corresponding expansion coefficient is zero or close to zero. The HMO method does not offer the possibility of considering a negative spin density, so that the prediction of a zero spin density is usually incorrect. Nonzero a_2 is supposed to be due the negative spin density, and thus it is necessary to use computations based on the method of configuration interaction or on the unrestricted SCF method; The McLachlan equation can also be used because it mimics the unrestricted MO procedure.

(b) The McLachlan equation for computing the spin density (see Appendix) has the form

$$\rho_\mu = c_\mu^2 + \lambda \sum_\nu c_\nu^2 \pi_{\mu\nu}$$

The expansion coefficients c_μ have the same meaning as in (a). $\pi_{\mu\nu}$ are atom – atom polarizabilities, and λ is a parameter which follows from the SCF theory.

$$\rho_1 = (0.707)^2 + 1.06 \left[(0.707)^2 \times 0.442 + (0.000) \times \right.$$

$$\times (-0.177) + (0.707)^2 (-0.265) \left. \right]$$

$$\rho_1 = 0.594$$

$$\rho_2 = 0.000 + 1.06 \left[(0.707)^2 (-0.177) + (0.000)(0.354) + \right.$$

$$+ (0.707)^2 (-0.177) \left. \right]$$

$$\rho_2 = -0.187$$

79. The plot below shows that there is a rough parallelism between the π-electron density on the chloro atom and the frequency of the NQR signal.

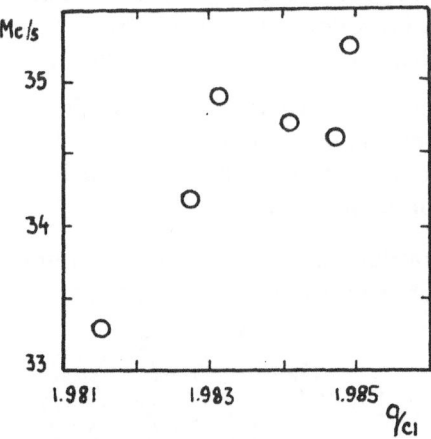

80. From Fig. 6 the approximate value of 1655 cm^{-1} is obtained.

81. (a) Product of the nucleophilic substitution (in position 2 the values of A_n and q are the smallest):

(b) Product of the radical substitution (in position 2 the value of A_r is the smallest and F is the highest):

(c) Product of the electrophilic substitution (in position 3 the value of A_e is the smallest and q is the highest):

82. The prediction based on atom localization energies for the electrophilic substitution is in agreement with experimental data because the smallest values of A_e are in positions 5 and 8. π-Electron densities, however, are not in agreement with these data, as the highest values are in positions 8 and 3.

Note: It is known that with electrophilic substitutions which take place only under drastic conditions there is a generally better agreement of experimental data with the localization energies than with π-electron densities.

83. It is well known that the free valence is a universal reactivity index with alternant hydrocarbons. Therefore, positions 5 and 8 are

expected to exhibit a maximum reactivity for both ionic and radical substitutions. Reaction with OsO_4: $5-6$; $7-8$; (the highest bond orders, the lowest values of ortho-localization energy). Reaction with maleic anhydride: $1 - 4$; $9 - 12$ (minimum para-localization energies).

84. Electrophilic: 1,3
    Nucleophilic: 4,8
    Radical: 4,8(1,3)

    The theory is in agreement with experiments.

85. (a) Experimentally it is very difficult to substitute a position which belongs to a bond characterized by a high bond order.

    (b) The value of the $C - C$ bond order is exceptionally high; therefore, the easy addition of bromine is very probable.

86. 5-Azauracil contains a π-electron system of α, β - unsaturated ketone. The high reactivity of these systems (RC-CH= CHR') is
    $$\overset{\|}{\underset{O}{}}$$
    well known. It is also known that the β-position is sensitive to nucleophilic agents (Michael's additions, hydration). A very low electron density in position 6 with 5-azauracil indicates that a nucleophilic attack will be easy and that even the attack of as weak a nucleophilic agent as water seems probable.

87.

   (a)   (b)

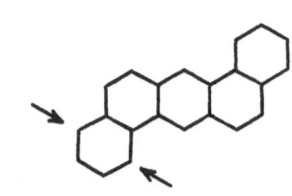

           30.879   30.879
          −27.366   −27.101
          ───────   ───────
            3.513   3.778

(c)

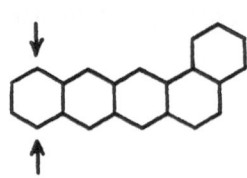

30.726
−27.101
――――――
3.625

(d)

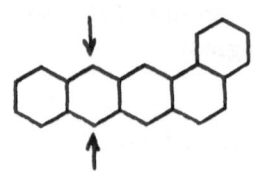

30.726
−27.449
――――――
3.277

(e)

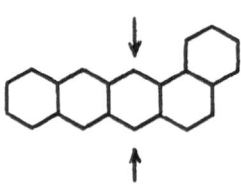

30.726
−27.366
――――――
3.360

(f)

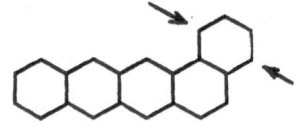

. 30.726
−26.931
――――――
3.795

(g)

28.222
−24.506
――――――
3.716

I: (a)
II: (d)
III: (g)

38. The predictions based on ortho–localization energies and bond orders agree with experimental data.

Note: The parallelity of ortho–localization energies and bond orders was found by R. D. Brown.[26]

89. 9–Chlorophenanthrene does not give a nucleophilic reaction with an OH⁻ ion; the energy appropriate to the activated complex is probably too high.

If an electronegative atom is introduced into the skeleton the reaction becomes possible.

Structure of the activated complex:

$$31\ a^2 = 1$$

$$a = \sqrt{\frac{1}{31}}$$

The largest excess of negative charge is created in position 10; therefore, the introduction of a nitrogen atom into this position should increase the reactivity most significantly.

90. Determination of the coefficients of a nonbonding molecular orbital (NBMO) according to Longuet–Higgins:

Normalization condition:

$$(3a)^2 + (-4a)^2 + (3a)^2 + (-3a)^2 + (2a)^2 + (-2a)^2 + (2a)^2 +$$
$$+ (-3a)^2 = 1$$

$$66a^2 = 1$$
$$16a^2 = 0.2424$$
$$9a^2 = 0.1363$$
$$4a^2 = 0.0606$$
$$a^2 = 0.0151$$

π-electron densities:

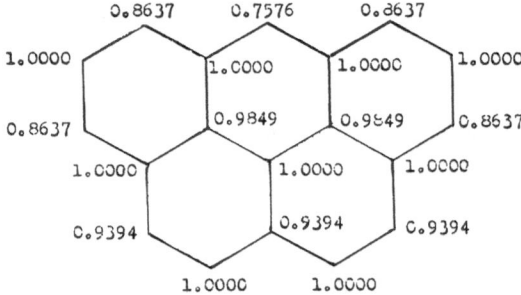

The center of nucleophilic reactivity in cation I will be in the position with the smallest electron density (0.7576).

91. Coupling reactions with phenols are carried out in an alkaline medium. A phenolate anion can be characterized by the Longuet-Hig-

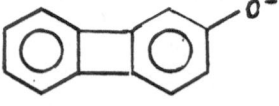

gins model. The procedure is the same as in Exercise 52.

Let us chose the value of the expansion coefficient in position 7 to be a; the value of the coefficient in positions 5 and 9 will then be - a, that in position 3, 2a, etc.

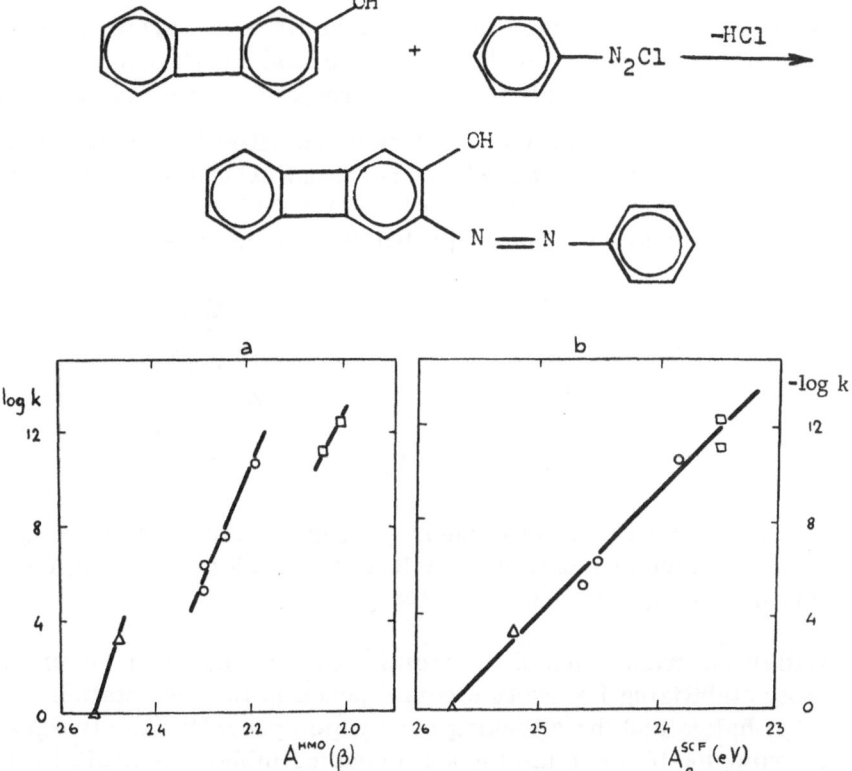

The largest π-electron density is in position 1.

In the coupling reaction the attacking agent is the ion $\text{C}_6\text{H}_5-\text{N}_2^+$.

The coupling is actually an electrophilic substitution, so that the following reaction scheme may be expected:

Fig. 16. Logarithms of bromination rates plotted against HMO (a) and SCF (b) electrophil localization energies. △ Benzene-like, ○ α-naphthalene-like, □ meso-anthracene-like positions.

92. Graphic treatment of the data (see **Fig. 16**) shows that there exist correlations between the experimental data and both series of theoretical data. However, in the first case (HMO values), the data are split (**Fig. 16a**) into three groups according to the types (classes) of substituted positions (benzene-like, α-naphthalene-like, meso-anthracene-like). This split is obviously not due to the steric hindrance because a single straight line results in the plot in which SCF localization energies are used (**Fig. 16b**). This shows the importance of the electronic repulsion term.

Graphic interpolations lead to the following predictions of the rate constant for position 1 in anthracene

$$\log k = 8.7 \quad \text{(from Fig. 16a)}$$
$$\log k = 8.0 \quad \text{(from Fig. 16b)}$$

93. In general the electronic interaction between two systems has a stabilizing effect only if the highest occupied (HO) orbital of a donor overlaps strongly the lowest free (LF) orbital of an acceptor.

(a) From the symmetry properties of the HOMO of the donor and from the fact that the 5s orbital is the LFAO of the acceptor it is apparent that in structure I the HOMO–LFAO overlap is zero. Hence, structure II is more probable, as confirmed experimentally.

(I)   (II)

(b) The interaction is of a donor – acceptor nature. In the anion (I) the overlap of frontier orbitals is zero, while in the cation (II) this overlap is efficient.

94. Within the framework of the frontier orbital theory of Fukui the most stabilizing interaction corresponds to that orientation of the alkyl halide and the attacking base which provides the strongest overlapping between the highest occupied molecular orbital (HOMO) of the base and the lowest free molecular orbital (LFMO) of the alkyl halide. As the HOMO of the attacking base is most frequently a nonbonding one-center localized orbital, the highest values in

squares of expansion coefficients on hydrogens in LFMO of the alkyl halide should determine the hydrogen atom to be abstracted.
The predicted favored positions are seen in figure to be in agreement with the Saytzeff rule.

95. On going from methylcyclopropane to the cycloprolylmethyl cation,
the $C_1 - C_2$ bond index is markedly reduced whereas the $C_2 - C_3$
bond index increases, which should result in an increase in the
1,2 bond length and in a decrease in the 2,3 bond length.

Owing to this change in bond lengths, in 2,3-bridged compounds
the strain energy would increase for the smaller bridging ring
and should lead to a small decrease in the rate of solvolysis. On
the other hand, with 1,2-bridged compounds a marked increase in
rate for the smaller bridging ring is observed. This finding can
be interpreted in an analogous way: here a considerable decrease
in the 1,2 bond index on going to the cation leads to a relief of strain
for the smaller ring.

96. The N_2H_2 molecule possesses 12 valence electrons, i.e., all orbital levels in the figure except the highest one are doubly occupied.
On rotation from cis to trans or from trans to cis through the nonplanar transition state, a crossing of the highest occupied and the
lowest free molecular orbital occurs. At the crossing point the
electrons cannot pass from one molecular orbital to the other
because it would mean a violation of the principle of conservation
of orbital symmetry postulated by Woodward and Hoffmann. Thus
the ground state of the cis isomer is not correlated with the ground
state of the trans isomer, and the cis – trans isomerization of N_2H_2
is therefore symmetry-forbidden.

This symmetry constraint disappears in H_2O_2, because here 14 valence electrons occupy all levels indicated in the figure. The cis
ground state correlates with the trans ground state and, therefore,
the concerted cis – trans isomerization of H_2O_2 is symmetry-allowed.

97. R e s u l t: 0.204 V; 0.405 V; 0.213 V.

First we compute the bicentric localization energies. In case I it
is the difference between the π-electron energy of benz[a]pyrene,

chrysene, and ethylene (Ia):

$$28.222 - 25.192 - 2.000 = 1.030$$

In case II the appropriate para-localization energy is given by the difference of the energy of dibenz[a,c]anthracene, phenanthrene, and benzene (IIa); the situation in the remaining case is clear. The para-localization energies have the following values: 3.494β and 3.178β. By substituting into the empirical equations the oxidation – reduction potentials mentioned above are obtained.

98. (a) Naphthalene:

$$E_{1/2} = 2.72 \ (-0.618) - 0.79 = -2.471 \ V$$

Anthracene:

$$E_{1/2} = 2.72 \ (-0.414) - 0.79 = -1.915 \ V$$

(b) Tetracene:

$$E_{1/2} = 2.72 \ (-0.295) - 0.79 = -1.592 \ V$$

Pentacene:

$$E_{1/2} = 2.72 \ (-0.220) - 0.79 = -1.388 \ V$$

The difference in half-wave potentials is approximately 0.2 V. One may expect, therefore, that the waves of tetracene and pentacene will be separated.

(c) With respect to the pairing properties of molecular orbitals of alternant hydrocarbons one may expect that all relationships mentioned will be linear:

$$|k_1| = |k_{-1}|$$

For the $N \rightarrow V_1$ excitation energy,

$$E(N \rightarrow V_1) = 2k_1 = 2 \left| k_{-1} \right|$$

(d) None.

99.

$$\vec{\mu}_\pi = \sum_i f_i \vec{r}_i ^*$$

This vector equation may be resolved into components:

$$\left| \vec{\mu}_\pi \right| = \sqrt{\mu_x^2 + \mu_y^2}$$

$$\mu_x = \sum_i f_i x_i$$

$$\mu_y = \sum_i f_i y_i$$

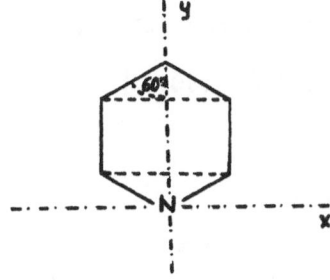

From the figure and from the knowledge of the bond lengths it is possible to compute the coordinates of the π-centers of pyridine.

---

*In this case i denotes the atom position and corresponds to the symbol μ that is usually used.

(a) Ground state:

net charges $(\xi_i = 1 - q_i)$

i	ξ_i
1	-0.195
2	0.077
3	-0.004
4	0.050

$$(\mu_\pi)_x = 4.80 \left[0 \times (-0.195) + 1.4 \times \sin 60° \times 0.077 - \right.$$

$$- 1.4 \times \sin 60° \times 0.077 + 1.4 \sin 60° (-0.004) -$$

$$\left. - 1.4 \sin 60° (-0.004) + 0 \times (0.050) \right]$$

$$(\mu_\pi)_x = 0$$

One could have expected a result like this, of course, because the pyridine molecule is symmetric with respect to the y axis. Only the y component will thus contribute to the total dipole moment:

$$\mu_\pi = 4.80 \left[0 \times (-0.195) + 2 \times 1.4 \cos 60° \times 0.077 - \right.$$

$$- 2(1.4 + 1.4 \cos 60°) 0.004 + (1.4 + 2 \times$$

$$\left. \times 1.4 \cos 60°) 0.050 \right]$$

$$\mu_\pi = 1.1 \text{ D}$$

(b) $1 \rightarrow 1'$ excited state:

net charges

i	ξ_i	i	ξ_i
1	-0.493	3	0.189
2	0.197	4	-0.271

$$\mu_\pi = 4.80 \left[0 \times (-0.493) + 2 \times 1.4 \cos 60° \times 0.197 + \right.$$

$$+ 2(1.4 + 1.4 \cos 60°) 0.189 - (1.4 + 2 \times 1.4 \cos 60°)$$

$$\left. \times 0.271 \right]$$

$$= 1.5 \text{ D}$$

Note: The total dipole moment is the sum of μ_π (π-electron contribution) and of μ_σ (σ-electron contribution).

00. We shall use Coulson's formula for computing the bond length:

$$x = 1.515 - \frac{0.180}{1 + 1.05 \left(\dfrac{1 - p}{p} \right)}$$

where p is the bond order.

Bond $1 - 2$

$$x = 1.515 - \frac{0.180}{1 + 1.05 \left(\dfrac{1 - 0.7386}{0.7386} \right)}$$

$$x = 1.38 \text{ Å}$$

Bond $2 - 3$

$$x = 1.515 - \frac{0.180}{1 + 1.05 \left(\dfrac{1 - 0.586}{0.586} \right)}$$

$$x = 1.41 \text{ Å}$$

Bond $11 - 12$

$$x = 1.515 - \frac{0.180}{1 + 1.05 \left(\dfrac{1 - 0.485}{0.485} \right)}$$

$$x = 1.43 \text{ Å}$$

101. The rule cannot be used: b, c, f, g, j, o.
The rule can be used conditionally: e(A), i(A), m, t(A), u.
The rule can be used: a(A), d, h, k(A), 1, n(A), p(A), q(A), r(A), s(A).

102. The rule cannot be used: a, b, i, o, p.
The following are aromatic: c, e, g, j, k, m.

103. There are 10 π-electrons present in the cation I, 12 in the anion II. In the cation I there is an unoccupied bonding orbital ($E = \alpha + 0.188\beta$) available (within the considered treatment), and one should, therefore, expect the cation I to display a high electron affinity and to be very easily reducible. From this aspect the hope for a synthesis of anion II is greater (although undoubtedly this anion would be comparatively easily oxidized).

104. According to the Pullmans' theory the necessary condition for a benzenoid hydrocarbon to be carcinogenic is the presence of a reactive K-sphere (the combined index of reactivity is smaller than or equal to 3.31β). The sufficient condition is the presence of a nonreactive L-sphere (or its absence) (the combined index of reactivity is larger than or equal to $5.60\beta^*$).

[Definition of quantities and terms: K-sphere − a pair of adjacent atoms in a bond, e.g., the 5 − 6 bond in benz[a]anthracene; L-sphere − a pair of atoms in the p-position, e.g., atoms 7 and 12 in the same molecule. The combined index of reactivity of the K-sphere (L-sphere): sum of the ortho-localization (para-localization) energy and the smaller of the localization energies of the atoms forming the K- (L-) sphere.]

L-sphere

Result:

Computed Values of the Combined Indices of Reactivity

Molecule	I	II	III	IV
K-sphere	−	3.245	3.303	3.274
L-sphere	5.020	5.401	5.644	−

---

* This fact concerns Wheland's localization energies[27]; the Pullmans characterized the reactivity by means of approximate values of localization energies (Dewar's reactivity numbers).

According to the criteria mentioned, it is expected that hydrocarbons III (3.30 < 3.31; 5.64 > 5.60) and IV (3.27 < 3.31; absence of L-sphere) are carcinogenic, hydrocarbons I and II are inactive. This agrees with experimental data.

III. Quantum – Chemical Data

1. HMO ORBITAL ENERGIES (IN β – UNITS)

BUTADIENE	HEXATRIENE	DECAPENTAENE	BENZENE	NAPHTHALENE
1.618	1.802	1.919	2.000	2.303
0.618	1.247	1.683	1.000	1.618
	0.445	1.310	1.000	1.303
		0.831		1.000
		0.285		0.618

PHENANTHRENE	ANTHRACENE	CHRYSENE	PYRENE	BENZANTHRACENE
2.435	2.414	2.499	2.532	2.485
1.950	2.000	2.167	2.000	2.176
1.516	1.414	1.701	1.802	1.755
1.306	1.414	1.540	1.347	1.480
1.142	1.000	1.286	1.247	1.323
0.769	1.000	1.216	1.000	1.166
0.605	0.414	0.875	0.879	1.000
		0.792	0.445	0.715
		0.520		0.452

TETRACENE	PENTACENE	CYCLOPENTADIENYL	TROPYL
2.467	2.496	2.000	2.000
2.194	2.303	0.618	1.247
1.777	2.000	0.618	1.247
1.467	1.618	–1.618	–0.445
1.294	1.496	–1.618	–0.445
1.194	1.303		–1.802
1.000	1.220		–1.802
0.777	1.000		
0.295	1.000		
	0.618		
	0.220		

METHYLENECYCLO-PROPENE	FULVENE	HEPTAFULVENE	PENTALENE	AZULENE
2.170	2.115	2.091	2.343	2.310
0.311	1.000	1.443	1.414	1.652
-1.000	0.618	1.247	1.000	1.356
-1.481	-0.254	0.216	0.471	0.887
	-1.618	-0.445	0.000	0.477
	-1.861	-0.776	-1.414	-0.400
		-1.302	-1.814	-0.738
		-1.974	-2.000	-1.579
				-1.869
				-2.095

HEPTALENE	FLUORANTHENE	PYRIDINE*	ANILINE†	QUINOLINE*
2.278	2.569	2.107	2.211	2.363
1.732	2.000	1.167	1.505	1.669
1.481	1.714	1.000	0.504	1.384
1.317	1.465	-0.841	-1.000	1.000
1.000	1.141	-1.000	-1.155	0.703
0.000	1.000	-1.934	-2.064	-0.527
-0.311	0.744			-1.000
-0.705	0.618			-1.232
-1.000	-0.371			-1.593
-1.732	-0.905			-2.267
-1.891	-1.000			
-2.170	-1.216			
	-1.497			
	-1.618			
	-2.268			
	-2.376			

$^*\delta_N = 0.5.$

$^†\delta_N = 1.0.$

ISOQUINOLINE[*]	1-AMINONAPHTHALENE[†]	2-AMINONAPHTHALENE[†]
2.340	2.400	2.369
1.733	1.810	1.915
1.319	1.500	1.372
1.078	1.000	1.219
0.646	1.000	0.758
−0.576	0.364	0.431
−0.920	−0.716	−0.651
−1.290	−1.000	−1.075
−1.549	−1.372	−1.317
−2.282	−1.653	−1.699
	−2.335	−2.322

2. HMO EXPANSION COEFFICIENTS

ALLYL

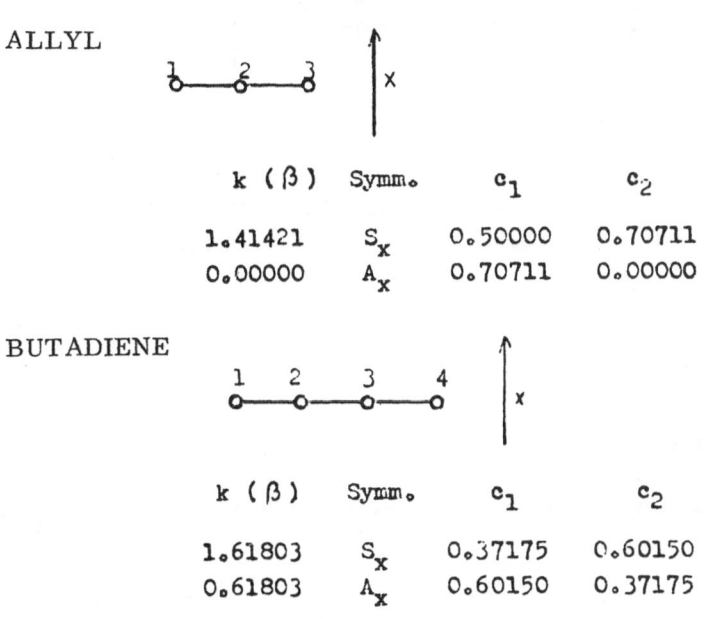

k (β)	Symm.	c_1	c_2
1.41421	S_x	0.50000	0.70711
0.00000	A_x	0.70711	0.00000

BUTADIENE

k (β)	Symm.	c_1	c_2
1.61803	S_x	0.37175	0.60150
0.61803	A_x	0.60150	0.37175

---

[*] $\delta_N = 0.5.$

[†] $\delta_N = 1.0.$

PENTADIENYL

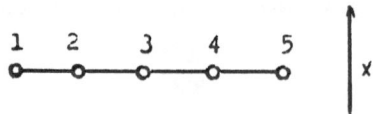

k (β)	Symm.	c_1	c_2	c_3
1.73205	S_x	0.28868	0.50000	0.57735
1.00000	A_x	0.50000	0.50000	0.00000
0.00000	S_x	0.57735	0.00000	-0.57735

HEXATRIENE

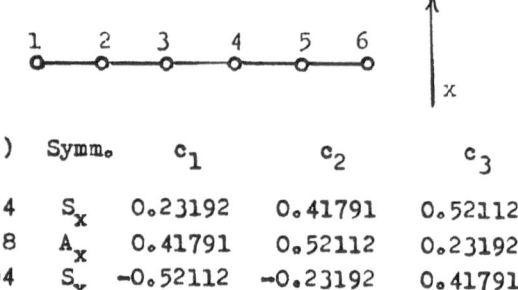

k (β)	Symm.	c_1	c_2	c_3
1.80194	S_x	0.23192	0.41791	0.52112
1.24698	A_x	0.41791	0.52112	0.23192
0.44504	S_x	-0.52112	-0.23192	0.41791

DECAPENTAENE

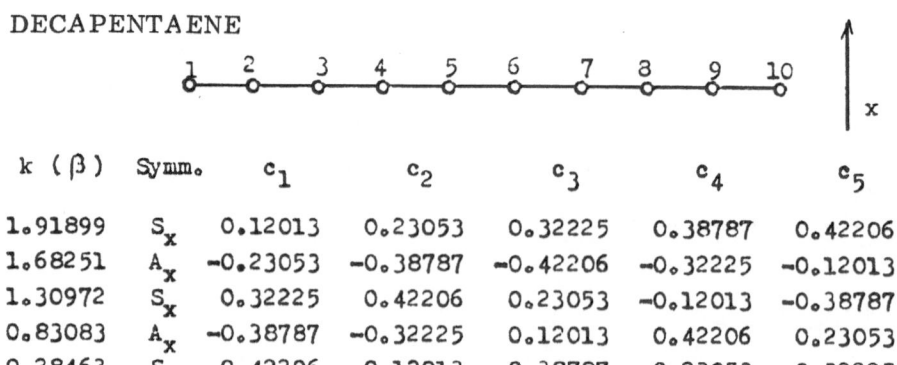

k (β)	Symm.	c_1	c_2	c_3	c_4	c_5
1.91899	S_x	0.12013	0.23053	0.32225	0.38787	0.42206
1.68251	A_x	-0.23053	-0.38787	-0.42206	-0.32225	-0.12013
1.30972	S_x	0.32225	0.42206	0.23053	-0.12013	-0.38787
0.83083	A_x	-0.38787	-0.32225	0.12013	0.42206	0.23053
0.28463	S_x	0.42206	0.12013	-0.38787	-0.23053	0.32225

BENZENE

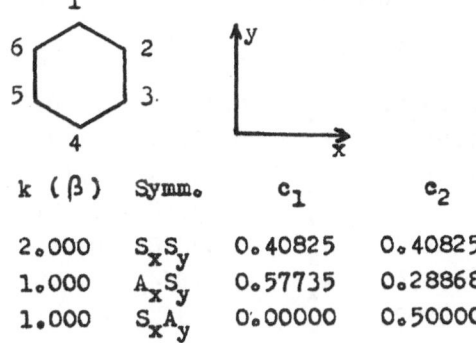

k (β)	Symm.	c_1	c_2
2.000	$S_x S_y$	0.40825	0.40825
1.000	$A_x S_y$	0.57735	0.28868
1.000	$S_x A_y$	0.00000	0.50000

NAPHTHALENE

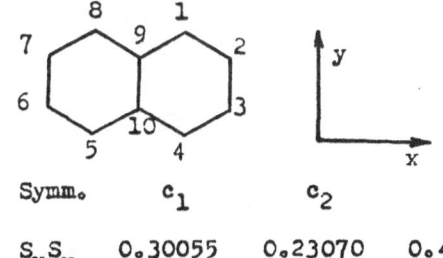

k (β)	Symm.	c_1	c_2	c_9
2.30278	$S_x S_y$	0.30055	0.23070	0.46140
1.61803	$S_x A_y$	0.26287	0.42533	0.00000
1.30278	$A_x S_y$	0.39958	0.17352	0.34705
1.00000	$S_x S_y$	0.00000	−0.40825	0.40825
0.61803	$A_x A_y$	0.42533	0.26286	0.00000

PHENANTHRENE

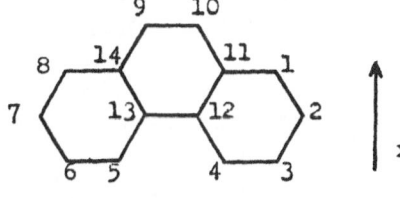

k (β)	Symm.	c_1	c_2	c_3	c_4	c_{10}
2.43476	S_x	0.20742	0.14974	0.15717	0.23293	0.24761
1.95063	A_x	0.32594	0.31427	0.28709	0.24573	0.10896
1.51627	S_x	−0.01954	0.20172	0.32540	0.29168	−0.44810
1.30580	S_x	−0.33337	−0.40581	−0.19654	0.14917	−0.09648
1.14238	A_x	−0.22129	0.14377	0.38554	0.29666	−0.18511
0.76905	A_x	−0.27301	−0.38859	−0.02584	0.36872	0.10098
0.60523	S_x	−0.34019	−0.04205	0.31474	0.23254	0.41503

PHENANTHRENE (continued)

k (β)	Symm.	c_{11}	c_{12}
2.43476	S_x	0.35527	0.40996
1.95063	A_x	0.32151	0.19224
1.51627	S_x	−0.23134	0.11686
1.30580	S_x	−0.02950	0.39132
1.14238	A_x	−0.39658	−0.04664
0.76905	A_x	0.17864	0.30941
0.60523	S_x	−0.16384	−0.17400

ANTHRACENE

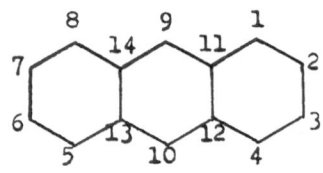

k (β)	Symm.	c_1	c_2	c_9	c_{11}
2.41421	$S_x S_y$	0.21488	0.15195	0.30389	0.36683
2.00000	$S_x A_y$	0.28868	0.28868	0.00000	0.28868
1.41421	$S_x S_y$	0.16828	0.40626	−0.23799	−0.16828
1.41421	$A_x S_y$	0.28076	0.11630	0.39706	0.28076
1.00000	$S_x A_y$	0.00000	0.35355	0.00000	−0.35355
1.00000	$A_x A_y$	0.40825	0.20412	0.00000	0.20412
0.41421	$A_x S_y$	0.31094	0.21987	−0.43973	−0.09108

TETRACENE

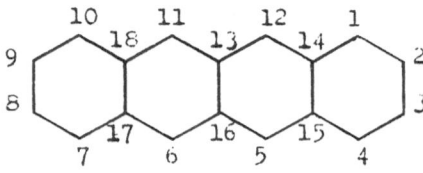

k (β)	Symm.	c_1	c_2	c_{12}	c_{13}	c_{14}
2.46673	$S_x S_y$	0.16052	0.10944	0.25972	0.35415	0.28651
2.19353	$S_x A_y$	0.25248	0.21154	0.15604	0.00000	0.34228
1.77748	$S_x S_y$	−0.23462	−0.30177	0.14501	0.37301	−0.11527
1.46673	$A_x S_y$	0.20816	0.08439	0.33682	0.27309	0.22093
1.29496	$S_x A_y$	−0.11322	−0.38385	0.18320	0.00000	0.23723
1.19353	$A_x A_y$	0.34228	0.15604	0.21154	0.00000	0.25248
1.00000	$S_x S_y$	0.00000	0.31623	0.00000	0.31623	−0.31623
0.77748	$A_x S_y$	−0.35476	−0.19958	0.21925	0.24670	−0.07623
0.29496	$A_x A_y$	−0.23723	−0.18320	0.38385	0.00000	0.11322

PYRENE

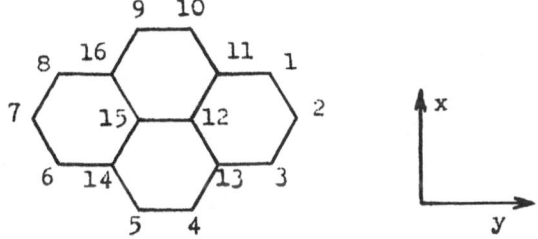

k (β)	Symm.	c_2	c_3	c_4	c_{12}	c_{13}
2.53209	$S_x S_y$	0.13769	0.17432	0.19823	0.39647	0.30371
2.00000	$A_x S_y$	0.29704	0.29704	0.09901	0.19803	0.29704
1.80194	$S_x A_y$	0.00000	0.16399	0.36849	0.00000	0.29550
1.34730	$S_x S_y$	0.46133	0.31077	−0.12273	−0.24546	−0.04262
1.24698	$A_x A_y$	0.00000	0.29550	0.16399	0.00000	0.36849
1.00000	$S_x S_y$	0.00000	0.00000	0.40825	−0.40825	0.00000
0.87939	$A_x S_y$	0.42424	0.18654	−0.13845	−0.27690	−0.26021
0.44504	$S_x A_y$	0.00000	0.36849	−0.29550	0.00000	0.16399

PERYLENE

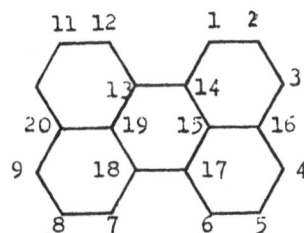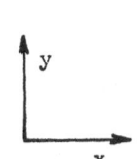

k (β)	Symm.	c_1	c_2	c_3	c_{14}	c_{15}	c_{16}
2.58836	$S_x S_y$	0.16958	0.11864	0.13751	0.32029	0.33916	0.23729
2.18194	$S_x A_y$	0.14810	0.18352	0.25232	0.13963	0.29620	0.26703
1.87939	$A_x S_y$	0.28868	0.21426	0.11401	0.32827	0.00000	0.00000
1.59358	$S_x S_y$	−0.04836	0.16734	0.31503	−0.24440	−0.09672	0.33468
1.53209	$A_x A_y$	0.28868	0.32827	0.21426	0.11401	0.00000	0.00000
1.00000	$S_x S_y$	0.28868	0.28868	0.00000	0.00000	−0.28868	−0.28868
1.00000	$S_x A_y$	0.25000	0.00000	−0.25000	0.25000	0.25000	−0.25000
1.00000	$S_x A_y$	0.22613	0.30151	0.07538	−0.07538	−0.37689	−0.22613
1.00000	$A_x S_y$	0.00000	0.28868	0.28868	−0.28868	0.00000	0.00000
0.34730	$A_x A_y$	−0.28868	0.11401	0.32827	−0.21426	0.00000	0.00000

FULVENE

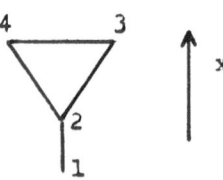

$k\ (\beta)$	Symm.	c_1	c_2	c_5	c_6
2.11492	S_x	0.42937	0.38513	0.52296	0.24727
1.00000	S_x	0.00000	−0.50000	0.50000	0.50000
0.61805	A_x	0.60150	0.37175	0.00000	0.00000
−0.25410	S_x	−0.35054	0.27952	−0.19044	0.74948
−1.61803	A_x	0.37175	−0.60150	0.00000	0.00000
−1.86081	S_x	0.43904	−0.15347	−0.66350	0.35657

METHYLENECYCLOPROPENE

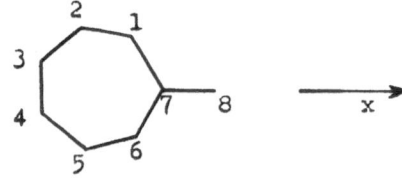

$k\ (\beta)$	Symm.	c_1	c_2	c_3
2.17009	S_x	0.28185	0.61163	0.52272
0.31111	S_x	−0.81522	−0.25362	0.36816
−1.00000	A_x	0.00000	0.00000	−0.70711
−1.48119	S_x	−0.50594	0.74939	−0.30201

HEPTAFULVENE

$k\ (\beta)$	Symm.	c_1	c_2	c_3	c_7	c_8
2.09118	S_x	0.39057	0.33247	0.30469	0.48428	0.23158
1.44272	S_x	−0.18000	0.22057	0.49822	−0.48026	−0.33288
1.24698	A_x	0.41791	0.52112	0.23192	0.00000	0.00000
0.21629	S_x	−0.33392	−0.22376	0.28552	0.15154	0.70061
−0.44504	A_x	−0.52112	0.23192	0.41791	0.00000	0.00000
−0.77638	S_x	−0.09978	0.46749	−0.26317	−0.39002	0.50236
−1.80194	A_x	0.23192	−0.41791	0.52112	0.00000	0.00000
−1.97382	S_x	0.44000	−0.26869	0.09035	−0.59979	0.30387

PENTALENE

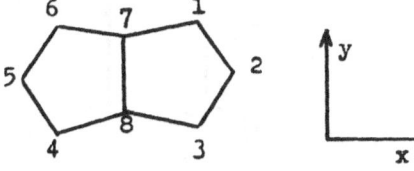

k (β)	Symm.	c_1	c_2	c_7
2.34292	$S_x S_y$	0.31793	0.27139	0.47349
1.41421	$S_x A_y$	0.35355	0.50000	0.00000
1.00000	$A_x S_y$	0.40825	0.00000	0.40825
0.47068	$S_x S_y$	-0.12068	-0.51279	0.45598

AZULENE

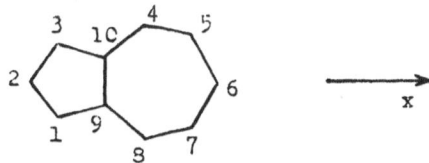

k (β)	Symm.	c_2	c_3	c_4	c_5	c_6	c_{10}
2.31028	S_x	0.27988	0.32330	0.28864	0.19981	0.17297	0.46703
1.65157	S_x	-0.32426	-0.26777	0.19090	0.43327	0.52467	-0.11798
1.35567	A_x	0.00000	0.22067	0.48406	0.35706	0.00000	0.29916
0.88697	S_x	0.58294	0.25853	-0.21856	0.15978	0.36028	-0.35363
0.47726	A_x	0.00000	-0.54282	0.16023	0.33548	0.00000	-0.25907

HEPTALENE

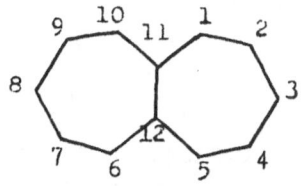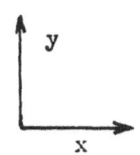

k (β)	Symm.	c_1	c_2	c_3	c_{11}
2.27841	$S_x S_y$	0.29211	0.20856	0.18307	0.45699
1.73205	$S_x A_y$	0.20412	0.35355	0.40825	0.00000
1.48119	$A_x S_y$	0.37469	0.25297	0.00000	0.30203
1.31743	$S_x S_y$	-0.05801	0.28906	0.43883	-0.36549
1.00000	$A_x A_y$	0.35355	0.35355	0.00000	0.00000
0.00000	$S_x A_y$	0.40825	0.00000	-0.40825	0.00000

PYRIDINE ($\delta_N = 0.5$)

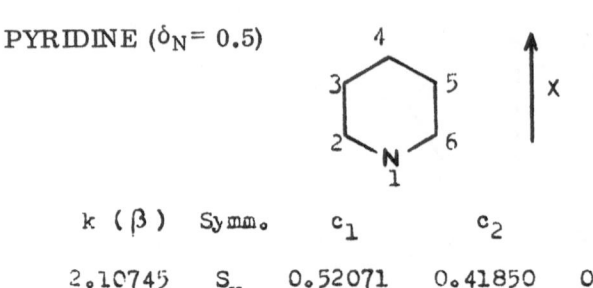

k (β)	Symm.	c_1	c_2	c_3	c_4
2.10745	S_x	0.52071	0.41850	0.36127	0.34285
1.16719	S_x	-0.57137	-0.19061	0.34890	0.59784
1.00000	A_x	0.00000	-0.50000	-0.50000	0.00000
-0.84096	S_x	-0.54591	0.36602	0.23810	-0.56626
-1.00000	A_x	0.00000	0.50000	-0.50000	0.00000
-1.93368	S_x	-0.32307	0.39313	-0.43711	0.45210

QUINOLINE ($\delta_N = 0.5$)

k (β)	c_1	c_2	c_3	c_4	c_5	c_6
2.36252	0.39474	0.26463	0.23046	0.27984	0.43066	0.26703
1.66893	0.37581	0.42457	0.33276	0.13079	-0.11449	-0.33658
1.38422	-0.39350	-0.06019	0.31019	0.48956	0.36747	0.30685
1.00000	0.00000	0.40825	0.40825	0.00000	-0.40825	0.00000
0.70328	-0.39513	-0.16592	0.27844	0.36174	-0.02403	-0.46424
-0.52710	-0.42220	0.35307	0.23610	-0.47752	0.01560	0.38873
-1.00000	0.00000	-0.40825	0.40825	0.00000	-0.40825	0.00000
-1.23157	-0.35025	0.22630	0.07155	-0.31442	0.31568	-0.45455
-1.59326	0.18655	-0.40901	0.46511	-0.33203	0.06390	0.21170
-2.26703	0.23715	-0.20509	0.22780	-0.31134	0.47802	-0.32123

QUINOLINE (continued)

k (β)	c_7	c_8	c_9	c_{10}
2.36252	0.20020	0.20595	0.28636	0.47058
1.66893	-0.44725	-0.40984	-0.23675	0.01473
1.38422	0.05728	-0.22756	-0.37228	-0.28775
1.00000	0.40825	0.40825	0.00000	-0.40825
0.70328	-0.30247	0.25152	0.47936	0.08960
-0.52710	-0.22050	-0.27250	0.36413	0.08057
-1.00000	0.40825	-0.40825	0.00000	0.40825
-1.23157	0.24413	0.15388	-0.43365	0.38018
-1.59326	-0.40120	0.42752	-0.27995	0.01851
-2.26703	0.25021	-0.24601	0.30750	-0.45111

ANILINE ($\delta_N = 1$)

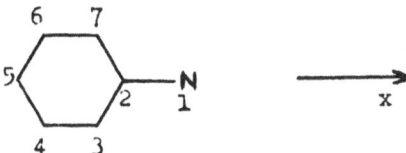

k (β)	Symm.	c_1	c_2	c_3	c_4	c_5
2.21051	S_x	0.44587	0.53972	0.37360	0.28612	0.25887
1.50467	S_x	-0.57215	-0.28875	0.06884	0.39233	0.52148
1.00000	A_x	0.00000	0.00000	-0.50000	-0.50000	0.00000
0.50428	S_x	-0.61959	0.30714	0.38724	-0.11186	-0.44365
-1.00000	A_x	0.00000	0.00000	0.50000	-0.50000	0.00000
-1.15538	S_x	-0.25710	0.55415	-0.19158	-0.33281	0.57610
-2.06408	S_x	0.15443	-0.47318	0.41113	-0.37542	0.36376

2-AMINONAPHTHALENE ($\delta_N = 1$)

k (β)	c_1	c_2	c_3	c_4	c_5	c_6
2.36919	0.33195	0.37178	0.27732	0.28526	0.39851	0.24420
1.91478	-0.14598	-0.47055	-0.24063	0.00980	0.25940	0.29586
1.37212	-0.14043	-0.10872	0.28341	0.49760	0.39935	0.13432
1.21887	0.35414	-0.06848	-0.12472	-0.08354	0.02290	-0.38869
0.75801	-0.11116	0.08623	0.53284	0.31767	-0.29204	-0.36855
0.43067	0.52559	0.28940	0.10736	-0.24316	-0.21209	0.21486

2-AMINONAPHTHALENE (continued)

k (β)	c_7	c_8	c_9	c_{10}	c_{11}
2.36919	0.18004	0.18236	0.25200	0.41468	0.27153
1.91478	0.30711	0.29218	0.25236	0.19103	-0.51439
1.37212	-0.21505	-0.42939	-0.37413	-0.08396	-0.29217
1.21887	-0.49666	-0.21668	0.23256	0.50014	-0.31290
0.75801	-0.01268	0.37816	0.27397	-0.17049	-0.35632
0.43067	0.30462	-0.08367	-0.34066	-0.06304	-0.50832

3. HMO ENERGY CHARACTERISTICS (IN β-UNITS)

Definitions:

n	number of π-electrons
m	number of C–C σ-bonds in the conjugated system
W	π-electron energy
DE/m	specific delocalization energy (delocalization energy per C–C bond)
k_2, k_1, k_{-1}, k_{-2}	energies of the frontier molecular orbital (two highest occupied and two lowest free)
$E(N \rightarrow V_1)$	energy of the $N \rightarrow V_1$ transition

Even Polyenes
$$C\!-\!(C\!-\!C)_{\overline{i}}\,C$$

i	n	m	W	DE/m	k_2	k_1	$E(N \rightarrow V_1)$
0	2	1	2.0000	0.0000	–	1.0000	2.0000
1	4	3	4.4721	0.1574	1.6180	0.6180	1.2360
2	6	5	6.9879	0.1976	1.2470	0.4450	0.8900
3	8	7	9.5175	0.2168	1.0000	0.3473	0.6946
4	10	9	12.0533	0.2281	0.8308	0.2846	0.5692
12	26	25	32.3969	0.2559	0.3473	0.1163	0.2326
17	36	35	45.1240	0.2607	0.2540	0.0849	0.1698

Odd Polyenes

$$C-(C-C)_i$$

i	m	W	DE/m	k_2	k_1	k_{-1}[a]	$E(N \rightarrow V_1)$
1	2	2.8284	0.4142	–	1.4142	0.0000	1.4142
2	4	5.4641	0.3660	1.7321	1.0000	0.0000	1.0000
3	6	8.0547	0.3425	1.4142	0.7654	0.0000	0.7654
4	8	10.6275	0.3284	1.1756	0.6180	0.0000	0.6180
5	10	13.1915	0.3192	1.0000	0.5176	0.0000	0.5176
12	24	31.0639	0.2943	0.4786	0.2411	0.0000	0.2411
17	34	43.8075	0.2885	0.3473	0.1743	0.0000	0.1743

[a] With the cation the φ_{-1} orbital (with energy k_{-1}) is unoccupied, with the anion it is doubly occupied.

Even Cyclopolyenes[a]

$$
\begin{array}{ccc}
C & - & C \\
| & & | \\
(C & - & C)_i
\end{array}
$$

i	Form	n	m	W	DE/m	k_2	k_1	$E(N \rightarrow V_1)$
1	–	4	4	4.0000	0.0000	2.0000	0.0000	–
1	diC[b]	2	4	4.0000	0.0000	–	2.0000	2.0000
2	–	6	6	8.0000	0.3333	1.0000	1.0000	2.0000
3	–	8	8	9.6569	0.2071	1.4142	0.0000	–
3	diC[b]	6	8	9.6569	0.2071	1.4142	1.4142	1.4142
4	–	10	10	12.9443	0.2944	0.6180	0.6180	1.2360
5	–	12	12	14.9282	0.2440	1.0000	0.0000	–
5	diC[b]	10	12	14.9282	0.2440	1.0000	1.0000	1.0000
6	–	14	14	17.9758	0.2840	0.4450	0.4450	0.8900

[a] For anti-Hückel cyclopolyenes (4n electrons) also the data of the corresponding dications are given. The values for the dianions can be determined easily from the data for the uncharged hydrocarbons.
[b] diC denotes dication.

Odd Cyclopolyenes

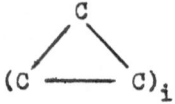

i	Form	n	m	W	DE/m	$k_1{}^a$	$k_{-1}{}^a$	$E(N \rightarrow V_1)$
1	C^b	2	3	4.0000	0.6667	2.0000	-1.0000	3.0000
2	A^b	6	5	6.4721	0.4944	0.6180	-1.6180	2.2360
3	C^b	6	7	8.9879	0.4268	1.2470	-0.4450	1.6920
4	A^b	10	9	11.5175	0.3908	0.3473	-1.0000	1.3473
5	C^b	10	11	14.0533	0.3685	0.8308	-0.2846	1.1154
6	A^b	14	13	16.5925	0.3533	0.2411	-0.7092	0.9503
7	C^b	14	15	19.1335	0.3422	0.6180	-0.2091	0.8271

[a] Twofold degenerate.   [b] C denotes cation, A denotes anion.

Benzenoid Hydrocarbons

Substance	n	m	W	DE/m	k_2	k_1	$E(N \rightarrow V_1)$
Benzene	6	6	8.0000	0.3333	1.0000	1.0000	2.0000
Naphthalene	10	11	13.6832	0.3348	1.0000	0.6180	1.2360
Anthracene	14	16	19.3137	0.3321	1.0000	0.4142	0.8284
Phenanthrene	14	16	19.4483	0.3405	0.7691	0.6052	1.2104
Tetracene	18	21	24.9308	0.3300	0.7775	0.2950	0.5900
Benz[a]anthracene	18	21	25.1012	0.3382	0.7150	0.4523	0.9046
Chrysene	18	21	25.1922	0.3425	0.7923	0.5201	1.0402
Benz[c]phenanthrene	18	21	25.1875	0.3423	0.6622	0.5676	1.1352
Triphenylene	18	21	25.3145	0.3483	0.6840	0.6840	1.3680
Pyrene	16	19	22.5055	0.3424	0.8794	0.4450	0.8900
Pentacene	22	26	30.5440	0.3286	0.6180	0.2197	0.4394
Benz[a]tetracene	22	26	30.7257	0.3356	0.6874	0.3271	0.6542
Pentaphene	22	26	30.7627	0.3370	0.5209	0.4372	0.8744
Benz[b]chrysene	22	26	30.8390	0.3400	0.7045	0.4048	0.8096
Dibenz[b,g]-phenanthrene	22	26	30.8340	0.3398	0.6601	0.4186	0.8372
Dibenz[a,c]-anthracene	22	26	30.9422	0.3439	0.7140	0.4991	0.9982
Dibenz[a,j]-anthracene	22	26	30.8792	0.3415	0.6180	0.4917	0.9834
Dibenz[a,h]-anthracene	22	26	30.8804	0.3416	0.6843	0.4735	0.9470

Nonalternant Hydrocarbons

System		n	m	W	DE/m	k_2	k_1	k_{-1}	k_{-2}	$E(N \to V_1)$
I Cyclopentadienyl	C	4		5.2362	0.2472	2.0000	0.6180	0.6180	-1.6180	—
	A	6	5	6.4722	0.4944	0.6180	0.6180	-1.6180	-1.6180	2.2360
II Indenyl	C	8		11.5808	0.3581	1.1935	0.7293	0.2950	-0.9016	0.4343
	A	10	10	12.1707	0.4171	0.7293	0.2950	-0.9016	-1.2950	1.1966
III Fluorenyl	C	12		17.5437	0.3696	1.0000	0.7046	0.1811	-0.8118	0.5236
	A	14	15	17.9059	0.3937	0.7046	0.1811	-0.8118	1.0000	0.9929
IV Tropyl	C	6		8.9879	0.4268	1.2470	1.2470	-0.4450	-0.4450	1.6920
	A	8	7	8.0979	0.2997	1.2470	-0.4450	-0.4450	-1.8019	—
V Benzotropyl	C	10		14.7040	0.3920	1.1557	0.8019	-0.2261	-0.5550	1.0231
	A	12	12	14.2518	0.3543	0.8019	-0.2261	-0.5550	-1.0818	0.3288
VI Dibenzo[a,c]tropyl	C	14		20.4664	0.3804	0.8503	0.7962	-0.1323	-0.6180	0.9285
	A	16	17	20.2018	0.3648	0.7962	-0.1323	-0.6180	-0.8932	0.4857
VII Dibenzo[a,d]tropyl	C	14		20.4222	0.3778	1.0000	0.6639	-0.1598	-0.5043	0.8237
	A	16	17	20.1026	0.3590	0.6639	-0.1598	-0.5043	-1.0000	0.3445
VIII Pentalene		8		10.4556	0.2728	1.0000	0.4707	0.0000	-1.4142	0.4707
VIIIa Dianion		10	9	10.4556	0.4951	0.4707	0.0000	-1.4142	-1.8136	1.4142
IX Azulene		10	11	13.3635	0.3058	0.8870	0.4773	-0.4004	-0.7376	0.8777
X Heptalene		12		15.6182	0.2783	1.0000	0.0000	-0.3111	-0.7046	0.3111
Xa Dication		10	13	15.6182	0.4322	1.3174	1.0000	0.0000	-0.3111	1.0000

4. HMO MOLECULAR DIAGRAMS

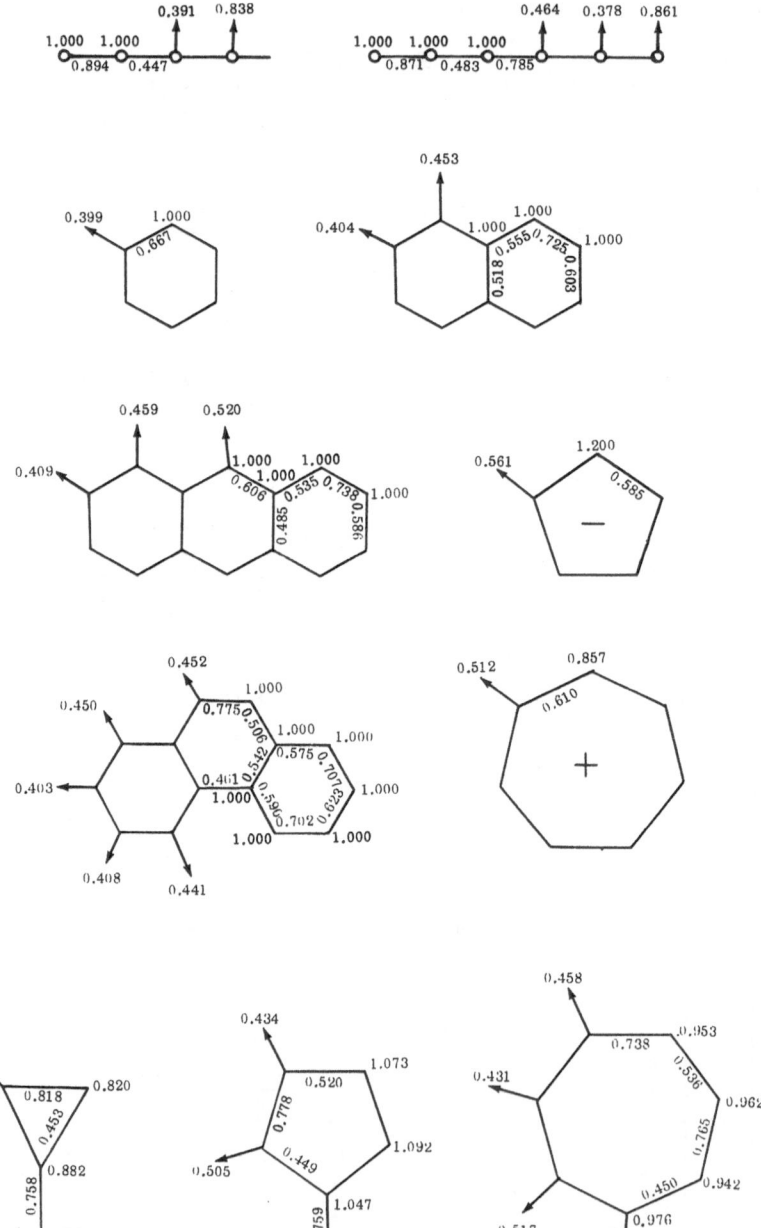

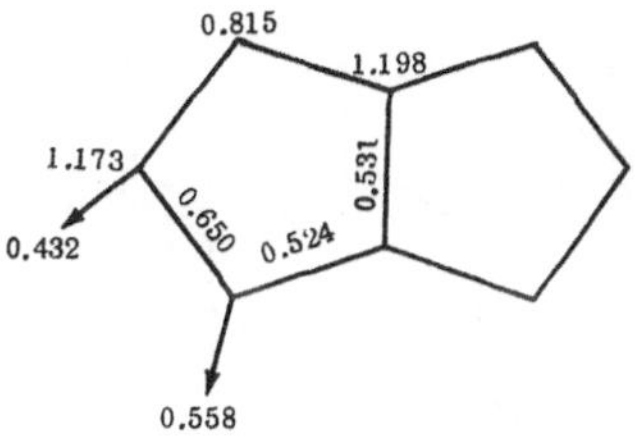
0.815
1.198
1.173
0.432
0.650
0.524
0.531
0.558

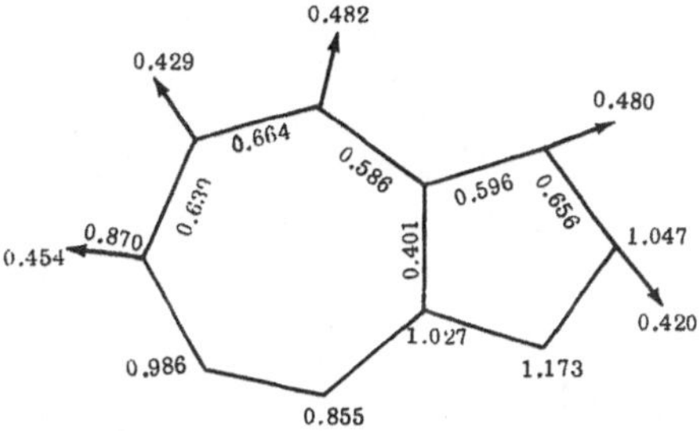
0.482
0.429
0.480
0.664
0.586
0.596
0.656
0.639
0.870
0.454
0.401
1.047
0.420
1.027
0.986
1.173
0.855

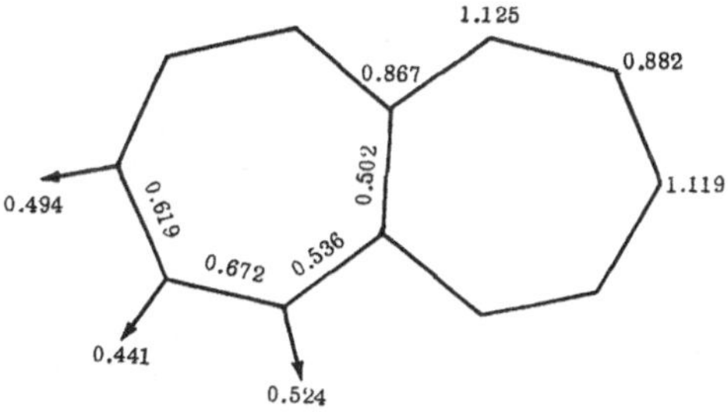
1.125
0.882
0.867
0.502
0.494
0.619
1.119
0.672
0.536
0.441
0.524

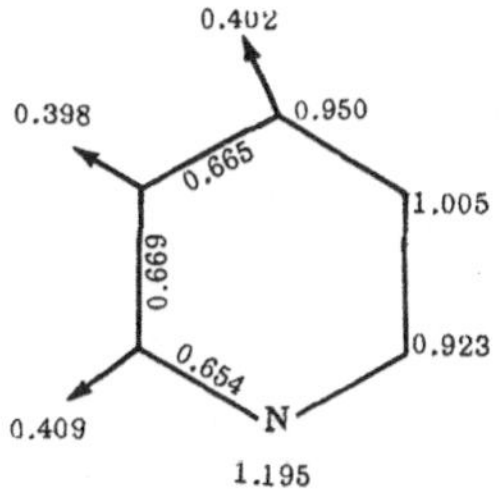
0.402
0.398
0.950
0.665
1.005
0.669
0.923
0.654
N
0.409
1.195

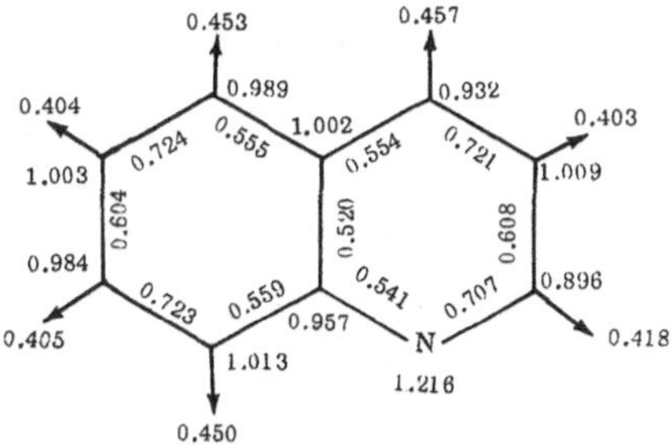
0.453
0.457
0.989
0.932
0.404
1.002
0.403
0.724
0.555
0.554
0.721
1.003
1.009
0.604
0.520
0.608
0.984
0.541
0.707
0.896
0.723
0.559
0.957
N
0.405
0.418
1.013
1.216
0.450

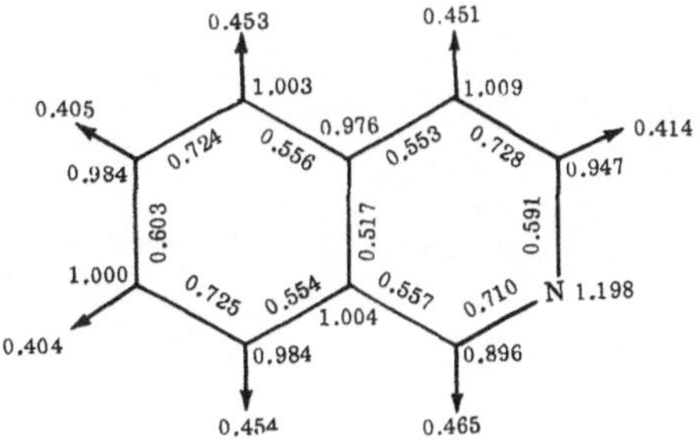
0.453
0.451
1.003
1.009
0.405
0.976
0.414
0.724
0.556
0.553
0.728
0.984
0.947
0.603
0.517
0.591
1.000
0.725
0.554
0.557
0.710
N 1.198
1.004
0.404
0.984
0.896
0.454
0.465

1.820
NH₂
0.431
0.450
0.601
0.938
1.089
0.681
0.657
0.997
0.394
1.072
0.418

0.458
0.512
0.404
1.025
1.130
1.814
0.724
0.550
0.907
0.569
0.651
0.437 NH₂
1.000
0.935
0.604
0.512
0.542
1.025
0.720
0.562
0.544
0.743
1.052
0.408
0.997
0.992
0.447
0.450
0.445

IV. Appendix

1. FORMULAS, SOME FUNDAMENTAL RELATIONS, AND DEFINITIONS

1. A. HMO Characteristics

Total π - electron energy:

$$W = \sum_{i=1}^{m} n_i E_i = \sum_{i=1}^{m} n_i \ (\alpha + k_i \beta) \tag{1}$$

Here m is the number of occupied molecular orbitals, n the number of electrons in the respective molecular orbital, (n = 1 or 2), and E the orbital energy.

Delocalization energy (resonance energy):

$$DE = W - m\, W_{Et} = W - m\, (2\alpha + 2\beta) \tag{2}$$

where m is the number of double bonds, and W_{Et}, the π-electron energy of ethylene.

Specific delocalization energy:

$$DE_n = \frac{DE}{n} \tag{3}$$

where n is the number of atoms.

$$DE_m = \frac{DE}{m} \tag{4}$$

where m is the number of C$-$C σ-bonds.

$\underline{N \to V_1 \text{ excitation energy:}}$

$$E(N \to V_1) = E_n - E_{n+1} = \alpha + k_n\beta - (\alpha + k_{n+1}\beta)$$

$$= (k_n - k_{n+1})\beta$$

$$\left[\text{or } E_1 - E_{-1} = (k_1 - k_{-1})\beta \right] \tag{5}$$

$\underline{\pi\text{-electron densities:}}$

$$q_\mu = \sum_{i=1}^{m} n_i\, c_{i\mu}^2 \tag{6}$$

where m is the number of occupied molecular orbitals.

$\underline{\pi\text{-charges}}$

$$\xi_\mu = n_\mu - q_\mu \tag{7}$$

where n_μ is the number of the electrons which the μ-th atom contributes to a conjugation.

$\underline{\text{Bond order:}}$

$$P_{\mu\nu} = \sum_{i=1}^{m} n_i\, c_{i\mu} c_{i\nu} \tag{8}$$

$\underline{\text{Free valence:}}$

$$F_\mu = N_{max} - \sum_{\substack{\text{neigh-} \\ \text{boring } \nu}} P_{\mu\nu} \tag{9}$$

For the sp^2 carbon atom $N_{max} = \sqrt{3}$.

$\underline{\text{Atom localization energy (Wheland):}}$

$$A_\mu = W - \sum_{T} W_T \tag{10}$$

where W_T is the π-electron energy of torso.

Superdelocalizabilities for electrophilic substitutions (Fukui et al.),
accurate (S), approximate (S'):

$$S_{\mu,e} = 2 \sum_{i=1}^{m} \frac{c_{i\mu}^2}{k_i} \quad ; \qquad S'_{\mu,e} = 2 \frac{c_{m\mu}^2}{k_m} \tag{11}$$

where m is the index of the highest occupied molecular orbital
(HOMO).

For nucleophilic substitution

$$S_{\mu,n} = -2 \sum_{l=m+1}^{n} \frac{c_{i\mu}^2}{k_i} \quad ; \qquad S'_{\mu,n} = -2 \frac{c_{(m+1)\mu}^2}{k_{m+1}} \tag{12}$$

where (m + 1) is the index of the lowest free molecular orbital
(LFMO).

For radical substitution

$$S_{\mu,r} = \tfrac{1}{2}(S_{\mu,e} + S_{\mu,n}) \quad ; \quad S'_{\mu,r} = \tfrac{1}{2}(S'_{\mu,e} + S'_{\mu,n}) \tag{13}$$

Polarizabilities (Coulson and Longuet-Higgins):

atom—atom

$$\pi_{\mu\nu} = 4 \sum_{j=1}^{m} \sum_{k=m+1}^{n} \frac{c_{j\mu}\, c_{j\nu}\, c_{k\mu}\, c_{k\nu}}{k_j - k_k} \tag{14}$$

bond—atom*

$$\pi_{\mu\nu,\rho} = 2 \sum_{j=1}^{m} \sum_{k=m+1}^{n} \frac{c_{j\rho}\, c_{k\rho}\,(c_{j\mu}\, c_{k\nu} + c_{j\nu}\, c_{k\mu})}{k_j - k_k} \tag{15}$$

$$\pi_{\mu\nu,\rho} = \tfrac{1}{2}\pi_{\rho,\mu\nu}$$

bond—bond

$$\pi_{\mu\nu,\rho\sigma} = 2 \sum_{j=1}^{m} \sum_{j=m+1}^{n} \frac{(c_{j\rho}\, c_{k\sigma} + c_{j\sigma}\, c_{k\rho})(c_{j\mu}c_{k\nu} + c_{j\nu}\, c_{k\mu})}{k_j - k_k}$$

*For alternant hydrocarbons the atom—bond and bond—atom polarizabilities are zero.

1.B. Relations between E, p, q, α, and β and Some Checks of the HMO Calculations

MO's as a linear combination of atomic orbitals:

$$\varphi_i = \sum_{\mu=1}^{n} c_{i\mu} \chi_\mu \tag{17}$$

From the Schrödinger equation follows directly (after multiplying by φ_i and integrating) for the orbital energy:

$$E_i = \frac{\int \varphi_i H \varphi_i \, d\tau}{\int \varphi_i^2 \, d\tau} \tag{18}$$

Using Equations (17) and (18) and applying variational method,

$$E_i = \sum_\mu c_{i\mu}^2 \alpha_\mu + 2 \sum_\mu \sum_{<\nu} c_{i\mu} c_{i\nu} \beta_{\mu\nu} \tag{19}$$

If the summation is carried out over the occupied MO's, the following equation is obtained:

$$W = \sum_{i=1}^{m} n_i E_i = \sum_{\mu=1}^{n} q_\mu \alpha_\mu + 2 \sum_\mu \sum_{<\nu} P_{\mu\nu} \beta_{\mu\nu} \tag{20}$$

Changes in π-electron densities and bond orders caused by changing the values of α and β:

$$\triangle q_\mu = \sum_\nu \pi_{\mu\nu} \triangle \alpha_\nu \tag{21}$$

$$\triangle q_\mu = \sum_{\rho\sigma} \pi_{\mu,\rho\sigma} \triangle \beta_{\rho\sigma} \tag{22}$$

$$\triangle P_{\mu\nu} = \sum_{\rho\sigma} \pi_{\mu\nu,\rho\sigma} \triangle \beta_{\rho\sigma} \tag{23}$$

$$\triangle P_{\mu\nu} = \sum_{\rho} \pi_{\mu\nu,\rho} \triangle \alpha_\rho \tag{24}$$

C h e c k :

$$\sum_{\nu} \pi_{\mu\nu} = 0 \qquad (25)$$

$$\sum_{\mu\nu} \pi_{\mu\nu,\rho} = 0 \qquad (26)$$

$$\sum_{\mu\nu} \pi_{\mu\nu,\rho\sigma} = 0 \qquad (27)$$

$$\sum_{\mu=1}^{n} q_{\mu} = \ell \quad \text{(total number of } \pi\text{-} \qquad (28)$$
$$\text{electrons)}$$

$$\sum_{\mu=1}^{n} F_{\mu} = n \sqrt{3} - 2 \sum_{i=1}^{m} k_i \qquad (29)$$

$$\sum_{\mu=1}^{n} S_{\mu,e} = 2 \sum_{occupied} \frac{1}{k_i} \qquad (30)$$

$$\sum_{\mu=1}^{n} S_{\mu,n} = -2 \sum_{unoccupied} \frac{1}{k_i} \qquad (31)$$

$$\sum_{\mu=1}^{n} S'_{\mu,e} = \frac{2}{k_m} \qquad (32)$$

$$\sum_{\mu=1}^{n} S'_{\mu,n} = - \frac{2}{k_{m+1}} \qquad (33)$$

$$\sum_{i=1}^{n} k_i = \sum_{\mu=1}^{n} a_{\mu\mu} \quad \text{(matrix trace)} \qquad (34)$$

$a_{\mu\mu}$ are diagonal matrix elements

$$\sum_{i=1}^{n} k_i^2 = \sum_{\mu} \sum_{\nu} a_{\mu\nu}^2 \qquad (35)$$

$a_{\mu\nu}$ are nondiagonal and diagonal matrix ($\mu = \nu$) elements; for hydrocarbons the sum of the squares of the matrix elements is equal to double the number of C–C bonds.

1.C.   A Special Method for Computing HMO Energies[28],[29]

$$\sum_{i=1}^{n} \frac{c_{i\mu}^2}{E - k_i} = \frac{E - \delta_\nu}{\rho_{\mu\nu}^2 + (E - \delta_\nu)\, \delta_\mu} \qquad (36)$$

$c_{i\mu}$ denotes expansion coefficients for the μ-th atomic orbital, where the substituent is bound. k_i is the energy of the i-th molecular orbital in the parent system in β units, E are the sought energies of the derivative under study, δ_ν and δ_μ denote the Coulomb integrals of the substituent and of the substituted atom, respectively. $\rho_{\mu\nu}$ is the resonance integral of the new $\mu\nu$-bond. The summation is carried out over all molecular orbitals.

1.D.   Transition Moment

For the $1 \rightarrow -1$ transition in the HMO method,

$$\vec{\lambda} = \sqrt{2} \sum_{\mu} c_{1\mu}\, c_{-1\mu}\, \vec{r} \qquad (37)$$

$c_{1\mu}$ ($c_{-1\mu}$) denote the expansion coefficient with the μ-th atomic orbital in the highest occupied (lowest free) molecular orbital.

1.E.   McLachlan Equation

This equation follows from the SCF theory preserving the different spin orbitals for α and β spins and is used for computing the spin densities:

$$\rho_\mu = c_{1\mu}^2 + \lambda \sum_{\nu} c_{i\nu}^2\, \pi_{\mu\nu} \qquad (38)$$

$c_{i\mu}$ is the HMO expansion coefficient with the μ-th atomic orbital in the molecular orbital φ_i which contains the unpaired electron; $\pi_{\mu\nu}$ are HMO atom − atom polarizabilities; λ is a parameter following from the SCF theory.

For radicals the sum of the spin densities is equal to 1, i.e., $\sum_{\mu} \rho_\mu = 1$, although ρ_μ can be negative.

2. ω TECHNIQUE AND RELATED METHODS[1,30]

If all the π-electron densities in the π-system are not unity, the Coulomb integrals do not have the same value. If for the position μ the relevant π-electron density is smaller than unity, the screening of the nuclear charge by the electrons is decreased. The carbon in the position μ, therefore, is more electronegative and the Coulomb integral α_μ should have a more negative value. Wheland and Mann[30] suggested the following equation for correcting Coulomb integrals:

$$\alpha_\mu = \alpha + (1 - q_\mu)\,\omega\,\beta \tag{39}$$

where ω is a dimensionless parameter, fitted empirically (for hydrocarbons $\omega = 1.4$). The Coulomb integrals obtained from Equation (39) can be used to compute new values of π-electron densities. This is continued until self-consistency is reached.

The Coulson and Golebiewski Method[31]

Just as the ω technique is used to correct the Coulomb integral with respect to the π-electron densities, the Coulson– Golebiewski method is used to correct the resonance integrals with respect to the bond orders. The matrix elements are defined as follows:

$$H_{\mu\mu} = \delta_\mu \tag{40}$$

$$H_{\mu\nu} = \exp\left[-2.683(0.120 - 0.180\,p_{\mu\nu})\right] \tag{41}$$

By constructing the matrix and solving the relevant secular problem, new bond orders are obtained. From these the new matrix elements are computed and this procedure is repeated until self-consistency is reached.

The Janssen-Sandström Method[32]

This SC procedure is similar to the ω-technique and to the Coulson– Golebiewski method. The matrix elements are defined as follows:

$$H_{\mu\mu} = \delta_\mu + \omega\,(n_\mu - q_\mu) \tag{42}$$

$$H_{\mu\nu} = \beta_{\mu\nu}\,(1 + 0.5\,p_{\mu\nu}) \tag{43}$$

$\omega = 1$, n_μ is the number of electrons in the atomic orbital μ.

3. VALENCE BOND METHOD (VB)[33]

For a given system all the possible structures whose wave functions
are linearly independent are written down. For example, for cyclic
systems all possible structures except those in which the bonds
intersect must be found. We obtain the so-called complete system
of canonical structures. Let this system contain N canonical structures and let ψ_K represent the wave function of structure K; then
the total wave function is sought in the following form:

$$\Psi = c_A \Psi_A + c_B \Psi_B + \ldots + c_N \Psi_N \tag{44}$$

The variation method leads to secular equations which have the
same form formally as the MO secular equations:

$$c_A (H_{AA} - S_{AA}E) + c_B (H_{AB} - S_{AB}E) + \ldots + c_N (H_{AN} - S_{AN}E) = 0$$

$$c_A (H_{BA} - S_{BA}E) + c_B (H_{BB} - S_{BB}E) + \ldots + c_N (H_{BN} - S_{BN}E) = 0$$

$$\vdots$$

$$c_A (H_{NA} - S_{NA}E) + c_B (H_{NB} - S_{NB}E) + \ldots + c_N (H_{NN} - S_{NN}E) = 0$$

$$\tag{45}$$

where

$$H_{KL} = \int \Psi_K H \Psi_L \, d\tau \quad ; \quad S_{KL} = \int \Psi_K \Psi_L \, d\tau \tag{46}$$

Secular determinant:

$$\begin{vmatrix} H_{AA} - S_{AA}E & H_{AB} - S_{AB}E & \cdots & H_{AN} - S_{AN}E \\ H_{BA} - S_{BA}E & H_{BB} - S_{BB}E & \cdots & H_{BN} - S_{BN}E \\ \vdots & & & \\ H_{NA} - S_{NA}E & H_{CB} - S_{CB}E & \cdots & H_{NN} - S_{NN}E \end{vmatrix} = 0 \tag{47}$$

Evaluation of the H_{KL} Integrals

It is first necessary to form so-called cycles. Let us start with the element H_{AA} for cyclobutadiene:

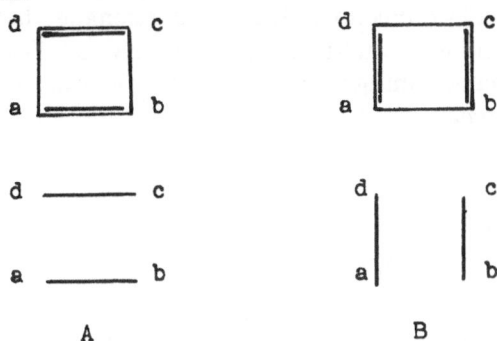

Spins are assigned to atomic orbitals. Let α be the spin assigned to a; then b must have spin β. We say that a and b form a cycle denoted by (a/b). In the numerator is the atomic orbital with α spin, in the denominator that with β spin. In a similar way the cycle (c/d) is formed. For H_{AA}, therefore, there are two cycles. For H_{AB} let us again assign α to a,

from which follows

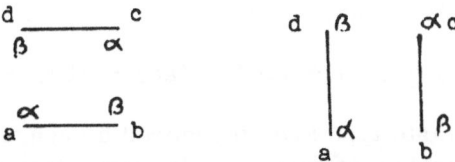

For H_{AB} there is thus only one cycle $(\frac{ac}{bd})$; in general,

$$H_{KL} = \frac{2^x}{2^y} \left[Q + \frac{3}{2} \left\{ \sum J \; (\alpha / \beta) - \sum J \; (\alpha, \beta) \right\} - \frac{1}{2} \sum J_{ij} \right] \quad (48)$$

where x is the number of cycles and y the number of bonds. $\Sigma J (\alpha/\beta)$ is the sum of single exchange integrals of the type $\int abcd\, H\, bacd\, d\tau$ (atomic orbitals, a, b, c, d) corresponding to the exchange between electrons in the same cycle with opposite spins, $\Sigma J(\alpha, \beta)$ is the sum of single exchange integrals corresponding to the exchange between electrons in the same cycle with the same spins, ΣJ_{ij} is the sum of all single exchange integrals, and Q is the Coulomb integral $\int abcd\, H\, abcd\, d\tau$.

For cyclobutadiene:

in integral H_{AA}:

$$x = 2, \quad y = 2$$

$$\Sigma J\ (\alpha/\beta) = \int abcd\, H\, bacd\, d\tau\ +\ \int abcd\, H\, abdc\, d\tau$$

or, for short,

$$\Sigma J\ (\alpha/\beta) = (ab) + (cd)$$

$$\Sigma J\ (\alpha, \beta) = 0$$

$$\Sigma J_{ij} = (ab) + (ac) + (ad) + (bc) + (bd) + (cd)$$

in integral H_{AB}:

$$x = 1, \quad y = 2$$

$$\Sigma J\ (\alpha/\beta) = (ab) + (ad) + (cb) + (cd)$$

$$\Sigma J\ (\alpha, \beta) = (ac) + (bd)$$

$$\Sigma J_{ij} = (ab) + (cd) + (ac) + (bd) + (ad) + (bc)$$

Integrals of the (ab) type are considered as nonzero only for adjacent atoms; if carbon atoms are being considered, all of them are put equal to the same value α (not to be confused with Hückel's α).

Evaluation of S_{KL} integrals

S_{KL} is equal to the coefficient with Q in expression (48) for H_{KL}.

4. FREE ELECTRON METHOD (FEMO)

In the free electron method (FEMO) the skeleton of the π-electron system in the molecule is considered, o be a potential box in which π-electrons move freely. If linear polyenes are being considered, the problem of a one-dimensional potential box must be solved. The relevant wave equation is

$$\frac{d^2 \psi}{dx^2} + \frac{8 \pi^2 m}{h^2} (E - V) \psi = 0 \tag{49}$$

Its solution is

$$\psi_n = \sqrt{\frac{2}{\ell}} \sin \frac{n \pi x}{\ell} \tag{50}$$

$$\tag{51}$$

$$E_n = \frac{n^2 h^2}{8m \ell^2} + V$$

where n is the quantum number, m the electron mass, and ℓ the length of box (chain). In the simplest case the potential due to the nuclei and the σ-electrons (i.e., the "core") is considered constant along the entire molecule. Thus it is possible to put $V = 0$.

$$E_n = \frac{n^2 h^2}{8m \ell^2} \tag{52}$$

5. SELF-CONSISTENT FIELD METHOD (SCF-MO-LCAO; POPLE[34])

According to Pople the matrix elements of the Hartree – Fock operator have the form

$$F_{\mu\mu} = I_\mu + \tfrac{1}{2} P_{\mu\mu} \gamma_{\mu\mu} + \sum_{\sigma(\neq\mu)} (F_{\sigma\sigma} - Z_\sigma) \gamma_{\mu\sigma} \tag{53}$$

$$F_{\mu\nu} = \beta_{\mu\nu}^{core} - \tfrac{1}{2} P_{\mu\nu} \gamma_{\mu\nu} \tag{54}$$

where I_μ is the valence state ionization potential and Z_σ is the charge of the core (for the sp^2 hybridized carbon atom $I_\mu = -11.22$ eV and $Z_\sigma = 1$); $\beta_{\mu\nu}$ is the resonance integral, which has a non–

zero value (-2.318 eV) only for the adjancent atoms; $P_{\mu\nu}$ denotes the bond order; and $\gamma_{\mu\nu}$ is the Coulomb electronic repulsion integral which can be approximated according to Mataga and Nishimoto.* If the both centers μ and ν are carbon atoms, the following formula is used:

$$\gamma_{\mu\nu} = \frac{14.399}{1.367 + r_{\mu\nu}} \tag{55}$$

Here $r_{\mu\nu}$ is the distance (in A) between the μ-th and ν-th $2p_z$ orbitals.

Since the elements of the F-matrix [Equations (53) and (54)] are functions of the coefficients, the SCF secular equations are solved by successive iterations. It is usual to start with HMO expansion coefficients, from which the F-matrix elements can be constructed. By solving the SCF secular equations

$$\sum_{\nu} c_{\nu} (F_{\mu\nu} - ES_{\mu\nu}) = 0 \qquad\qquad \mu = 1, 2, \ldots \tag{56}$$

a new set of coefficients is obtained, and from these the new F-matrix elements are determined. This is repeated until the coefficients from the (n + 1)-th and n-th iteration no longer differ, which means the self-consistency has been reached.

Total SCF π-Electron Energy[34]

Employing a procedure similar to that for obtaining the expression for the total HMO π-electron energy, we get the total SCF π-electron energy in the form

$$\mathcal{E}_{\pi} = \sum_{\mu} P_{\mu\mu} (I_{\mu} + \tfrac{1}{4} P_{\mu\mu} \gamma_{\mu\mu}) + 2\sum_{\mu<\nu}\sum P_{\mu\nu} \beta_{\mu\nu}^{core} +$$

$$+ \sum_{\mu<\nu}\sum \left\{ (P_{\mu\mu} - Z_{\mu})(P_{\nu\nu} - Z_{\nu}) - \tfrac{1}{2} P_{\mu\nu}^2 \right\} \gamma_{\mu\nu} - \tag{57}$$

$$- \sum_{\mu<\nu}\sum Z_{\mu} Z_{\nu} \gamma_{\mu\nu}$$

---

*Repulsion integrals can also be approximated in other ways, e.g., according to Pariser and Parr. For singlet excited states, however, the Mataga approximation seems to be most suitable.

where the symbols used have the same meaning as in Equations (53) and (54). For alternant hydrocarbons the expression can be simplified:

$$\mathcal{E}_\pi = \sum_\mu (I_\mu + \tfrac{1}{4} \gamma_{\mu\mu}) + 2 \sum_\mu \sum_{<\nu} P_{\mu\nu} \, \beta_{\mu\nu}^{core}$$

$$- \tfrac{1}{2} \sum_\mu \sum_{<\nu} P_{\mu\nu}^2 \, \gamma_{\mu\nu} - \sum_\mu \sum_{<\nu} \gamma_{\mu\nu} \tag{58}$$

If the terms connected with electronic repulsion ($\gamma_{\mu\nu}$) are neglected, the well-known expression for the total HMO π-electron energy (20) is obtained.

6. LIMITED CONFIGURATION INTERACTION METHOD (LCI-MO-LCAO, PARISER, PARR[35])

The LCI wave function has the form

$$\phi = c_1 \bar{\Psi}_1 + c_2 \bar{\Psi}_2 + \ldots + c_n \bar{\Psi}_n \tag{59}$$

where $\bar{\psi}_i$ are wave functions of singlet configurations (Slater's determinants); V_0 means the ground state (for the singly excited singlet the symbol $^1\psi_{i \to j}$ or $|1, i \to j >$ is used). (The number gives the multiplicity of the state under consideration, and the symbol $i \to j$ indicates that the configuration is due to the excitation of an electron from the i-th MO to the j-th MO.) The variational method, as with HMO or SCF-MO, leads to a secular determinant.

In the Pariser – Parr and Pople approximation the following expressions hold:

for diagonal matrix elements of the ground state:

$$< V_0 | H | V_0 > = 0 \tag{60}$$

for a matrix element between the ground state and the singly excited singlet:

$$< V_0 | H | 1, i \to j > = \sqrt{2} \, F_{ij} \tag{61}$$

for a matrix element between two singly excited singlets:

$$\langle 1, i \to j | H | 1, k \to \ell \rangle = F_{j\ell}\delta_{ik} - F_{ik}\delta_{j\ell} +$$

$$2(kj|G|\ell i) - (kj|G|i\ell) \tag{62}$$

where

$$F_{j\ell} = \sum_{\mu}\sum_{\nu} c_{j\mu} F_{\mu\nu} c_{\ell\nu} \tag{63}$$

$$(kj|G|\ell i) = \sum_{\mu}\sum_{\nu} c_{k\mu} c_{\ell\mu} c_{j\nu} c_{i\nu} \gamma_{\mu\nu} \tag{64}$$

$$F_{\mu\nu} = \beta_{\mu\nu} - 0.5\, P_{\mu\nu} \gamma_{\mu\nu} \tag{65}$$

If both centers μ and ν are carbon atoms, $\gamma_{\mu\nu}$ can be expressed as follows:

$$\gamma_{\mu\nu} = \frac{14.399}{1.367 + r_{\mu\nu}} \tag{66}$$

In Equations (62)–(66) $c_{j\mu}$ is the expansion coefficient with the μ-th atomic orbital in the j-th molecular orbital, $\beta_{\mu\nu}$ is the resonance integral, which has a nonzero value $(-2.318\ eV)$ only for the adjacent atoms, $p_{\mu\nu}$ indicates the HMO bond order, $\gamma_{\mu\nu}$ is the Coulomb electronic repulsion integral, and $r_{\mu\nu}$ is the distance (Å) between the μ-th and the ν-th $2p_z$ orbitals (usually 1.40 Å).

7. UNITS

1 eV corresponds to 23,063 cal/mol
1 eV corresponds to 8.068 kcm^{-1}
1 kcm^{-1} corresponds to 1.2394×10^{-1} eV
$h = 6.624 \times 10^{-27}$ erg sec (Planck's constant)
$k = 1.3805 \times 10^{-16}$ erg deg^{-1} (Boltzmann's constant)

V. References

1a. A. Streitwieser, Jr., "Molecular Orbital Theory for Organic Chemists," John Wiley & Sons, New York, 1961.

1b. M. J. S. Dewar, "The Molecular Orbital Theory of Organic Chemistry," McGraw-Hill, New York, 1969.

2. J. D. Roberts, "Notes on Molecular Orbital Calculations," Benjamin, New York, 1961.

3. J. N. Murrell, "The Theory of the Electronic Spectra of Organic Molecules," Methuen & Co., London, and John Wiley & Sons, New York, 1963.

4. J. N. Murrell, S. F. A. Kettle, and J. M. Tedder, "Valence Theory," John Wiley & Sons, New York, 1965.

5. A. Julg and O. Julg. "Exercises de chimie quantique," Dunod, Paris, 1967.

6. E. Heilbronner and H. Bock, "Das HMO-Modell und seine Anwendung," Verlag Chemie, Weinheim, 1968.

7. H. C. Longuet-Higgins, Trans. Faraday Soc. 45:173 (1949).

8. R. Hoffman, J. Chem. Phys. 39:1397 (1963).

9. R. S. Mulliken, C. A. Rieke, D. Orloff, and H. Orloff, J. Chem. Phys. 17:1248 (1949).

10. K. Hafner and W. Bauer, Angew. Chem. Internat. Ed. 7:297 (1968).

11. Ref. 3, Chapter 8.

12. E. Heilbronner, J. Michl. J.-P. Weber, and R. Zahradník, Theoret. Chim. Acta 6:141 (1966).

13. R. W. Fessenden and R. H. Schuler, J. Chem. Phys. 39:2147 (1963).

14. E. A. C. Lucken, "Methods in Heterocyclic Chemistry" (ed. A. R. Katritzky), p. 89, Academic Press, New York, 1963.

15. G. Berthier, B. Pullman, and J. Pontis, J. Chim. Phys. 49:367 (1952).

16. H. C. Longuet-Higgins, J. Chem. Phys. 18:265 (1950).

17. H. C. Longuet-Higgins, Proc. Chem. Soc. 1957:157.

18. L. Altschuler and E. Berliner, J. Am. Chem. Soc. 88:5837 (1966).

19. J. Koutecký, R. Zahradník, and J. Arient, Collection Czech. Chem. Commun. 27:2490 (1962).

20. G. J. Hoijtink, Rec. Trav. Chim. Pays-Bas, 74:1525 (1955).

21. A. T. Watson and F. A. Matsen, J. Chem. Phys., 18:1305 (1950).

22. Tables of Interatomic Distances and Configuration in Molecules and Ions, Special Publication No. 11, The Chemical Society, London, 1958.

23. A. Pullman and B. Pullman, "Cancérisation par les substances chimiques et structure moléculaire," Masson, Paris, 1955.

24. J. Koutecký, J. Paldus, and R. Zahradník, J. Chem. Phys. 36:3129 (1962).

25. A. Carrington, Quart. Revs. 17:67 (1963).

26.   R. D. Brown, J. Chem. Soc., 1950:3249.

27.   J. Koutecký and R. Zahrandník, Cancer Research 21:457 (1961).

28.   J.Koutecký, Trans. Faraday Soc. 55:505 (1959).

29.   M. J. S. Dewar, Proc. Cambr. Phil. Soc. 45:638 (1949).

30.   G. W. Wheland and D. E. Mann, J. Chem. Phys. 17:264 (1949).

31.   C. A. Coulson and A. Golebiewski, Proc. Phys. Soc. 78:1310 (1961).

32.   M. J. Jansen and J. Sandström, Tetrahedron 20:2339 (1964).

33.   This part is based on the description given in S. Glasstone, "Theoretical Chemis-
      try," Van Nostrand, New York, 1944.

34.   J. A. Pople, Trans. Faraday Soc. 49:1375 (1953).

35.   R. Pariser and R. G. Parr, J. Chem. Phys. 21:466 (1953).